I0038425

Molecular Biology of the Gene

Molecular Biology of the Gene

Curtis Holmes

New York

Published by Syrawood Publishing House,
750 Third Avenue, 9th Floor,
New York, NY 10017, USA
www.syrawoodpublishinghouse.com

Molecular Biology of the Gene
Curtis Holmes

International Standard Book Number: 978-1-68286-823-2 (Hardback)

Cataloging-in-Publication Data

Molecular biology of the gene / Curtis Holmes.
p. cm.
Includes bibliographical references and index.
ISBN 978-1-68286-823-2
1. Molecular biology. 2. Molecular genetics. I. Holmes, Curtis.
QH506 .M65 2019
572.8--dc23

TABLE OF CONTENTS

PREFACE

A gene is a sequence of DNA or RNA that codes for a molecule that has a unique function. During gene expression, the DNA is copied into RNA. The transmission of genes to the next generation is the basis of inheritance of phenotypic traits. The study of the structure and function of genes at the molecular level is approached from the discipline of molecular genetics, which is a branch of molecular biology. It explores the aspects of heredity, variation and mutation by studying chromosomes and gene expression. The understanding of gene amplification techniques, particularly polymerase chain reaction and molecular cloning, separation and detection of DNA and mRNA, etc. are vital to the understanding of the molecular biology of genes. This book aims to shed light on some of the unexplored aspects of this area of study. Some of the diverse topics covered herein address the significant aspects of molecular biology of the gene. In this book, constant effort has been made to make the understanding of the difficult concepts, as easy and informative as possible, for the readers.

Given below is the chapter wise description of the book:

Chapter 1- A gene is a DNA or RNA sequence, which codes for a specific molecule to deliver a particular function. The study of genes, variation and heredity in living organisms is called genetics. This is an introductory chapter, which will provide a brief introduction of gene and the study of genetics. Some of the topics covered in this chapter cover the branches of genetics and the molecular basis of inheritance.

Chapter 2- The genetic material of an organism is known as the genome. It is made up of DNA. The genes, mitochondrial DNA, noncoding DNA and chloroplast DNA together make up the genome. This chapter has been carefully written to provide an easy understanding of the basic concepts of genome size and complexity and coding regions. It also explores in depth the genome in the case of eukaryotic and prokaryotic organisms as well as in viruses.

Chapter 3- DNA or deoxyribonucleic acid is a double helix structure that carries the genetic instruction useful for the growth, functioning and reproduction of organisms. RNA or ribonucleic acid is a molecule essential for the regulation, coding, decoding and expression of gene. The aim of this chapter is to provide an easy understanding of DNA and RNA. It elucidates topics like DNA condensation, DNA fragmentation, DNA methylation, RNA interference, RNA processing, etc.

Chapter 4- The process of determining the order of nucleotides within a DNA molecule is known as DNA sequencing. This involves determining the order of adenine, cytosine, guanine and thymine in a DNA molecule. The topics elaborated in this chapter on DNA sequencing theory, next generation sequencing, whole genome sequencing, exome sequencing, polony sequencing, etc. will provide a holistic understanding of DNA sequencing.

Chapter 5- DNA replication is a fundamental process of biological inheritance. It refers to the production of two replicas of DNA from a DNA molecule. This chapter discusses in detail the processes involved in DNA replication, such as polymerase chain reaction and

DNA replication in prokaryotic and eukaryotic organisms.

Chapter 6- Gene expression is a process, which involves the synthesis of a gene product such as a functional protein or functional RNA from the information derived from a gene. All the diverse aspects of gene expression have been carefully analyzed in this chapter, such as genetic code, antitermination, artificial gene synthesis, gene silencing, RNA splicing, etc.

Chapter 7- Any change in the sequence composition of cellular molecules like DNA, RNA and proteins over multiple generations is referred to as molecular evolution. Some of the major topics of study in this domain include the genetic basis of speciation, origin of new genes, natural selection vs. neutral evolution, genetic nature of complex traits, etc. This chapter delves into the fundamentals of mutation, gene duplication and molecular evolution.

At the end, I would like to thank all those who dedicated their time and efforts for the successful completion of this book. I also wish to convey my gratitude towards my friends and family who supported me at every step.

Curtis Holmes

Chapter 1

An Introduction to Gene and Genetics

A gene is a DNA or RNA sequence, which codes for a specific molecule to deliver a particular function. The study of genes, variation and heredity in living organisms is called genetics. This is an introductory chapter, which will provide a brief introduction of gene and the study of genetics. Some of the topics covered in this chapter cover the branches of genetics and the molecular basis of inheritance.

Gene

All living beings have genes. They exist throughout the body. Genes are a set of instructions that determine what the organism is like, its appearance, how it survives, and how it behaves in its environment.

Genes are made of a substance called deoxyribonucleic acid or DNA. They give instructions for a living being to make molecules called proteins.

Genes are a section of DNA that is in charge of different functions like making proteins. Long strands of DNA with lots of genes make up chromosomes. DNA molecules are found in chromosomes. Chromosomes are located inside of the nucleus of cells.

Each chromosome is one long single molecule of DNA. This DNA contains important genetic information.

Chromosomes have a unique structure, which helps to keep the DNA tightly wrapped around the proteins called histones. If the DNA molecules were not bound by the histones, they would be too long to fit inside of the cell.

Genes vary in complexity. In humans, they range in size from a few hundred DNA bases to more than 2 million bases.

Different living things have different shapes and numbers of chromosomes. Humans have 23 pairs of chromosomes, or a total of 46. A donkey has 31 pairs of chromosomes, a hedgehog has 44, and a fruit fly has just 4.

DNA contains the biological instructions that make each species unique.

DNA is passed from adult organisms to their offspring during reproduction. The building blocks of DNA are called nucleotides. Nucleotides have three parts: A phosphate group, a sugar group and one of four types of nitrogen bases.

Combination

A gene consists of a long combination of four different nucleotide bases, or chemicals. There are many possible combinations.

The four nucleotides are:

- A (adenine);
- C (cytosine);
- G (guanine);
- T (thymine).

Different combinations of the letters ACGT give people different characteristics. For example, a person with the combination ATCGTT may have blue eyes, while somebody with the combination ATCGCT may have brown eyes.

Genes decide almost everything about a living being. One or more genes can affect a specific trait. Genes may interact with an individual's environment too and change what the gene makes.

Genes affect hundreds of internal and external factors, such as whether a person will get a particular color of eyes or what diseases they may develop.

Some diseases, such as sickle-cell anemia and Huntington's disease, are inherited, and these are also affected by genes.

Molecular Basis

A gene is a basic unit of heredity in a living organism. Genes come from our parents. We may inherit our physical traits and the likelihood of getting certain diseases and conditions from a parent.

Genes contain the data needed to build and maintain cells and pass genetic information to offspring.

Each cell contains two sets of chromosomes: One set comes from the mother and the other comes from the father. The male sperm and the female egg carry a single set of 23 chromosomes each, including 22 autosomes plus an X or Y sex chromosome.

A female inherits an X chromosome from each parent, but a male inherits an X chromosome from their mother and a Y chromosome from their father.

Genetic Testing

Genetic tests can show an individual whether they have a genetic risk for a specific disease. The results can help healthcare professionals diagnose conditions.

The HGP is expected to speed up progress in medicine, but there is still much to learn, especially regarding how genes behave and how they can be used in treatment. At least 350 biotechnology-based products are currently in clinical trials.

In 2005, the HapMap, a catalog of common genetic variation or haplotypes in the human genome, was created. This data has helped to speed up the search for the genes involved in common human diseases.

The Epigenome

In recent years, geneticists have found another layer of heritable genetic data that is not held in the genome, but in the "epigenome," a group of chemical compounds that can tell the genome what to do.

In the body, DNA holds the instructions for building proteins, and these proteins are responsible for a number of functions in a cell.

The epigenome is made up of chemical compounds and proteins that can attach to DNA and direct a variety of actions. These actions include turning genes on and off. This can control the production of proteins in particular cells.

Gene switches can turn genes on and off at different times and for different lengths of time.

Recently, scientists have discovered genetic switches that increase the lifespan and boost fitness in worms. They believe these could be linked to an increased lifespan in mammals.

The genetic switches that they have discovered involve enzymes that are ramped up after mild stress during early development.

This increase in enzyme production continues to affect the expression of genes throughout the animal's life.

This could lead to a breakthrough in the goal to develop drugs that can flip these switches to improve human metabolic function and increase longevity.

Gene Marking

When epigenomic compounds attach themselves to DNA in the cell and modify the function, they are said to have "marked" the genome.

The marks do not change the sequence of the DNA, but they do change the way cells use the DNA's instructions.

The marks can be passed on from cell to cell as they divide, and they can even be passed from one generation to the next.

Specialized cells can control many functions in the body. For example, specialized cells in red blood cells make proteins that carry oxygen from air to the rest of the body. The epigenome controls many of these changes within the genome.

The chemical tags on the DNA and histones can become rearranged as the specialized cells and the epigenome change throughout a person's lifetime.

Lifestyle and environmental factors such as smoking, diet and infectious diseases can bring about changes in the epigenome. They can expose a person to pressures that prompt chemical responses.

These responses can lead to direct changes in the epigenome, and some of these changes can be damaging. Some human diseases are due to malfunctions in the proteins that "read" and "write" epigenomic marks.

Some of these changes are linked to the development of disease.

Cancer can result from changes in the genome, the epigenome or both. Changes in the epigenome can switch on or off the genes that are involved in cell growth or the immune response. These changes can cause uncontrolled growth, a feature of cancer, or a failure of the immune system to destroy tumors.

Researchers in The Cancer Genome Atlas (TCGA) network are comparing the genomes and epigenomes of normal cells with those of cancer cells in the hope of compiling a current and complete list of possible epigenomic changes that can lead to cancer.

Researchers in epigenomics are focused on trying to chart the locations and understand the functions of all the chemical tags that mark the genome. This information may lead to a better understanding of the human body and knowledge of ways to improve human health.

Gene Therapy

In gene therapy, genes are inserted into a patient's cells and tissues to treat a disease, usually a hereditary disease. Gene therapy uses sections of DNA to treat or prevent disease. This science is still in its early stages, but there has been some success.

For example, in 2016, scientists reported that they had managed to improve the eyesight of 3 adult patients with congenital blindness by using gene therapy.

In 2017, a reproductive endocrinologist, named John Zhang, and a team at the New Hope Fertility Center in New York used a technique called mitochondrial replacement therapy in a revolutionary way.

They announced the birth of a child to a mother carrying a fatal genetic defect. Researchers combined DNA from two women and one man to bypass the defect.

The result was a healthy baby boy with three genetic parents. This type of research is still in the early stages, and much is still unknown, but results look promising.

Scientists are looking at different ways of treating cancer using gene therapy. Experimental gene therapy may use patients' own blood cells to kills cancer cells. In one study, 82 percent of patients had their cancer shrink by at least half at some point during treatment.

Gene Testing to Predict Cancer

Another use of genetic information is to help predict who is likely to develop a disease, for example, early-onset Alzheimer's disease and breast cancer.

Women with the BRCA1 gene have a significantly higher chance of developing breast cancer. A woman can have a test to find out whether she carries that gene. BRCA1 carriers have a 50 percent chance of passing the anomaly to each of their children.

Genetic Tests for Personalized Therapy

Scientists say that one day we will be able to test a patient to find out which specific medicines are best for them, depending on their genetic makeup. Some medicines work well for some patients, but not for others. Gene therapy is still a growing science, but in time, it may become a viable medical treatment.

Types of Gene

House Keeping Genes

They are those genes which are constantly expressing themselves in a cell because their products are required for the normal cellular activities, e.g., genes for glycolysis, ATP-ase.

Non-constitutive Genes

The genes are not always expressing themselves in a cell. They are switched on or off according to the requirement of cellular activities, e.g., gene for nitrate reductase in plants, lactose system in Escherichia coli. Non- constitutive genes are of further two types, inducible and repressible.

Inducible Genes

The genes are switched on in response to the presence of a chemical substance or inducer which is required for the functioning of the product of gene activity, e.g., nitrate for nitrate reductase.

Repressible Genes

They are those genes which continue to express themselves till a chemical (often an end product) inhibits or represses their activity. Inhibition by an end product is known as feedback repression.

Multigenes

It is a group of similar or nearly similar genes for meeting requirement of time and tissue specific products, e.g., globin gene family (e, 5, (3, y on chromosome 11, oc and 8 on chromosome 16).

Repeated Genes

The genes occur in multiple copies because their products are required in larger quantity, e.g., histone genes, tRNA genes, rRNA genes, actin genes.

Single Copy Genes

The genes are present in single copies (occasionally 2—3 times), e.g., protein coding genes. They form 60—70% of the functional genes. Duplications, mutations and exon reshuffling can form new genes.

Pseudogenes

They are genes which have homology to functional genes but are unable to produce functional products due to intervening nonsense codons, insertions, deletions and inactivation of promoter regions, e.g., several of snRNA genes.

Processed Genes

They are eukaryotic genes which lack introns. Processed genes have been formed probably due to reverse transcription or retroviruses. Processed genes are generally non-functional as they lack promoters.

Split Genes

They were discovered in 1977 by many workers but credit is given to Sharp and Roberts (1977). Split genes are those genes which possess extra or nonessential regions interspersed with essential or coding parts. The nonessential parts are called introns, spacer DNA or intervening sequences (IVS). Essential or coding parts are called exons. Transcribed intronic regions are removed before RNA passes out into cytoplasm. Split genes are characteristic of eukaryotes.

However, certain eukaryotic genes are completely exonic or non-split e.g., histone genes, interferon genes. Split genes have also been recorded in prokaryotes, thymidylate synthase gene and ribonucleotide reductase gene in T_4. A gene that produces calcitonin in thyroid forms a neuropeptide in hypothalamus by removing an exon. Adenovirus has also a mechanism to produce 15—20 different proteins from a single transcriptional unit by differential splicing.

Transposons

They are segments of DNA that can jump or move from one place in the genome to another. Transposons were first discovered by Me Clintock (1951) in case of Maize when she found that a segment of DNA moved into gene coding for pigmented kernels and produced light colored kernels.

Transposons possess repetitive DNA, either similar or inverted, at their ends, some 5, 7 or 9-nucleotide long. Enzyme transposase separates the segment from its original by cleaving the repetitive sequences at its ends.

There are many types of transposons. In human beings the most common types of transposons belong to Alu family (having a site for cutting by restriction enzyme Alu I). The number of nucleotides per transposon is about 300 with about 300,000 copies in the genome. Passage of transposons from one place to another brings about reshuffling of nucleotide sequences in genes. Reshuffling in introns often changes expression of genes, e.g., proto-oncogenes → oncogenes. New genes may develop by exon shuffling. Other changes caused by transposons are mutations, through insertions, deletions and translocations.

Overlapping Genes

Overlapping genes are defined as a pair of adjacent genes whose coding regions are partially overlapping. In other words, a single stretch of DNA codes for portions of two separate proteins. Such an arrangement of genetic code is ubiquitous. Many overlapping genes have been identified in the genomes of prokaryotes, eukaryotes, mitochondria, and viruses.

For two genes to overlap, the signal to begin transcription for one must reside inside the second gene, whose transcriptional start site is further "upstream." In addition, the "stop" signal for the second gene must not be read by the ribosome during translation, using the RNA copy of the gene. This is possible because RNA is read in triplets, meaning that it can contain three separate sequences that can be "read" by the cell's protein-making machinery. Such sequences of nucleotide triplets are called reading frames, and they are different in the RNA transcripts of the overlapping genes.

Overlapping genes enable the production of more proteins from a given region of DNA than is possible if the genes were arranged sequentially. Indeed, for the bacteriophage PhiX174, overlapping of genes is necessary. The amount of DNA present in the circular, single-stranded DNA genome of this virus would not be sufficient to encode the eleven bacteriophage proteins if transcription occurred in a linear fashion, one gene after another.

The genome economy afforded by overlapping genes extends to the human genome. The recently completed sequencing of the human genome has revealed between 30,000 and 70,000 genes. Yet evidence suggests that the human genome encodes 100,000 to 200,000 proteins. At least part of the information for the extra proteins may come from the presence of hitherto undiscovered overlapping genes, although more may come from alternative splicing of exons in a single gene.

In algae called *Guillardia*, a structure called a nucleomorph contains only about 500,000 base pairs of DNA, a very small genome, yet produces almost 500 proteins. Part of the efficient packaging of the genome is due to 44 overlapping genes. A nucleomorph is a remnant of a nucleus from an ancient eukaryotic organism that became incorporated into the algae.

One consequence of overlapping genes is to reduce the tolerance for mutation. Virtual experiments conducted within the past several years using a software system called Avida have indicated that overlapping reduces the probability of accumulating so-called neutral mutations in a gene (mutations that have no effect). Neutral mutations are unlikely with overlapping genes, because the mutation must have no effect on two genes with different reading frames.

The evolutionary origin of overlapping genes is not yet clear. Recent research indicates that they may have arisen due to several mutational events. These may include the loss of a signal to stop the transcription process in a gene and a shift in the reading sequence of the genetic components.

Structural Genes

Structural genes are those genes which have encoded information for the synthesis of chemical substances required for cellular machinery.

The chemical substances may be:

a. Polypeptides for the formation of structural proteins (e.g., colloidal complex of protoplasm, cell membranes, elastin of ligaments, collagen of tendons or cartilage, actin of muscles, tubulin of microtubules, etc.).

b. Polypeptides for the synthesis of enzymes.

c. Transport proteins like haemoglobin of erythrocytes, lipid transporting proteins, carrier proteins of cell membranes, etc.

d. Proteinaceous hormones, e.g., insulin, growth hormone, parathyroid hormone.

e. Antibodies, antigens, certain toxins, blood coagulation factors, etc.

f. Non-translated RNAs like tRNAs, rRNA. Broadly speaking, structural genes either produce mRNAs for synthesis of polypeptides/proteins/enzymes or non-coding RNAs.

Regulatory Genes

Regulatory genes do not transcribe RNAs for controlling structure and functioning of the cells. Instead, they control the functions of structural genes. The important regulatory genes are promoters, terminators, operators and repressor producing or regulator genes. Repressor does not take part in cellular activity. Instead, it regulates the activity of other genes. Therefore, repressor producing gene is of intermediate nature.

Tissue Specific Genes

They are genes which are expressed only in certain specific tissues and not in others.

Genetics

Genetics is a field of biology that studies how traits are passed from parents to their offspring. The passing of traits from parents to offspring is known as heredity, therefore, genetics is the study of heredity. This introduction to genetics takes you through the basic components of genetics such as DNA, genes, chromosomes and genetic inheritance.

Genetics is built around molecules called DNA. DNA molecules hold all the genetic information for an organism. It provides cells with the information they need to perform tasks that allow an organism to grow, survive and reproduce. A gene is one particular section of a DNA molecule that tells a cell to perform one specific task.

Heredity is what makes children look like their parents. During reproduction, DNA is replicated and passed from a parent to their offspring. This inheritance of genetic material by offspring influences the appearance and behavior of the offspring. The environment that an organism lives in can also influence how genes are expressed.

DNA

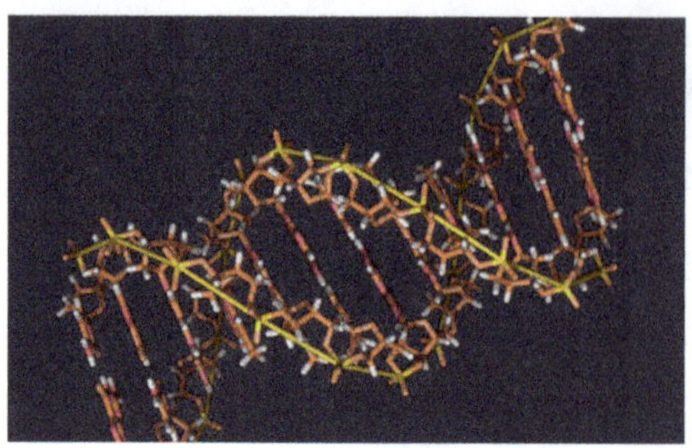

DNA is the cornerstone of genetics and is the perfect place to start for an introduction to genetics. DNA stands for deoxyribonucleic acid and it is the molecule that holds the genetic information for a cell and an organism.

A DNA molecule contains a code that can be used by a cell to express certain genes. Specific sections of a DNA molecule provide the information to build specific proteins which can then be used by a cell to express the desired gene.

A DNA molecule is a nucleic acid, one of the four molecules of life. It comes in the form of a long, linear molecule referred to as a strand. Each strand of DNA is bonded to a second strand of DNA to form a DNA double helix. In eukaryotic cells, DNA is found in the nucleus as a tightly coiled double helix.

DNA molecules are replicated during cell division. When a cell divides, the two new cells contain all the same DNA that the original cell had.

In sexual reproduction with two parents, half of the DNA of the offspring is provided by each of the parents. The genetic material of a child is made from 50% of their mother's DNA and 50% their father's DNA.

Genes

A gene is a specific segment of a DNA molecule that holds the information for one specific protein. DNA molecules have a unique code for each gene which codes for their specific protein. Some organisms can have more than 100,000 different genes so they will have 100,000 unique sequences of DNA 'code'.

Genes are the basic unit of heredity. The genes of an individual are determined by their parent or parents. A bacteria that is born by one parent cell splitting into two cells and has the exact same genes as their one parent cell.

A human, on the other hand, has two copies of each gene – one set from their mother and a second set from their father. Different forms of the same gene are called alleles. For each gene, a human can have two different alleles or two of the same alleles – one from each parent.

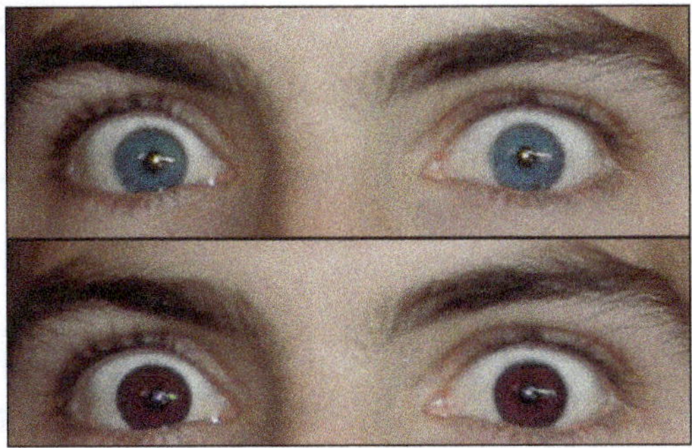

Physical traits such as eye color or height are often determined by the combination of multiple genes. The environment an individual lives in also impacts how genes are expressed.

Chromosomes

A chromosome is a structure made from tightly packed strands of DNA and proteins called histones. Strands of DNA are tightly wrapped around the histone proteins and form into long worm-shaped structures called 'chromatids'. Two chromatids join together to form a chromosome.

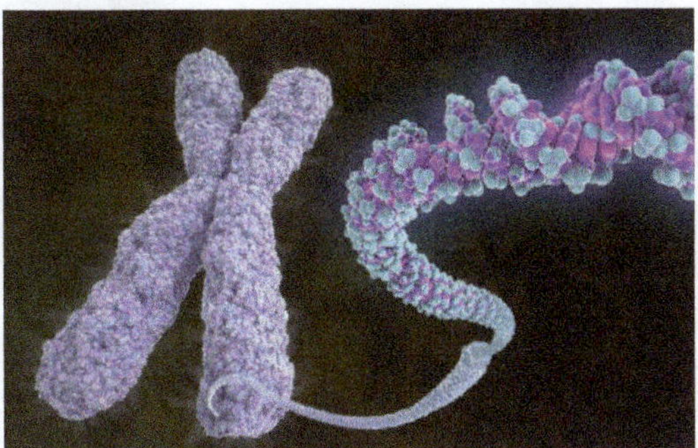

Chromosomes are formed in the nucleus of a cell when a cell is dividing. It is possible to see chromosomes under an ordinary light microscope if the cell is in the right stage of cell division.

The number of chromosomes varies between species. Humans have 46 chromosomes. Some species can have many more than 100 chromosomes while others can have as little as two.

Genetic Inheritance

Inheritance is the backbone of genetics and is an important topic to cover in an introduction to genetics. Long before DNA had been discovered and the word 'genetics' had been invented people were studying the inheritance of traits from one generation to the next.

Genetic inheritance occurs both in sexual reproduction and asexual reproduction. In sexual reproduction, two organisms contribute DNA to produce a new organism. In asexual reproduction, one organism provides all DNA and produces a clone of themselves. In either, genetic material is passed from one generation to the next.

Branches of Genetics

Classical genetics, which remains the foundation for all other areas in genetics, is concerned primarily with the method by which genetic traits—classified as dominant(always expressed), recessive (subordinate to a dominant trait), intermediate (partially expressed), or polygenic (due to multiple genes)—are transmitted in plants and animals. These traits may be sex-linked (resulting from the action of a gene on the sex, or X, chromosome) or autosomal (resulting from the action of a gene on a chromosome other than a sex chromosome). Classical genetics began with Mendel's study of inheritance in garden peas and continues with studies of inheritance in many different plants and animals. Today a prime reason for performing classical genetics is for gene discovery—the finding and assembling of a set of genes that affects a biological property of interest.

Cytogenetics

Cytogenetics, the microscopic study of chromosomes, blends the skills of cytologists, who study the structure and activities of cells, with those of geneticists, who study genes. Cytologists discovered chromosomes and the way in which they duplicate and separate during cell division at about the same time that geneticist began to understand the behavior of genes at the cellular level. The close correlation between the two disciplines led to their combination.

Plant cytogenetics early became an important subdivision of cytogenetics because, as a general rule, plant chromosomes are larger than those of animals. Animal cytogenetics became important after the development of the so-called squash technique, in which entire cells are pressed flat on a piece of glass and observed through a microscope; the human chromosomes were numbered using this technique.

Today there are multiple ways to attach molecular labels to specific genes and chromosomes, as well as to specific RNAs and proteins that make these molecules easily discernible from other components of cells, thereby greatly facilitating cytogenetics research.

Microbial Genetics

Microorganisms were generally ignored by the early geneticists because they are small in size and were thought to lack variable traits and the sexual reproduction necessary for a mixing of genes from different organisms. After it was discovered that microorganisms have many different physical and physiological characteristics that are amenable to study, they became objects of great interest to geneticists because of their small size and the fact that they reproduce much more rapidly than larger organisms. Bacteria became important model organisms in genetic analysis, and many discoveries of general interest in genetics arose from their study. Bacterial genetics is the centre of cloning technology.

Viral genetics is another key part of microbial genetics. The genetics of viruses that attack bacteria were the first to be elucidated. Since then, studies and findings of viral genetics have been applied to viruses pathogenic on plants and animals, including humans. Viruses are also used as vectors (agents that carry and introduce modified genetic material into an organism) in DNA technology.

Molecular Genetics

Molecular genetics is the study of the molecular structure of DNA, its cellular activities (including its replication), and its influence in determining the overall makeup of an organism. Molecular genetics relies heavily on genetic engineering (recombinant DNA technology), which can be used to modify organisms by adding foreign DNA, thereby forming transgenic organisms. Since the early 1980s, these techniques have been used extensively in basic biological research and are also fundamental to the biotechnology industry, which is devoted to the manufacture of agricultural and medical products. Trans genes is forms the basis of gene therapy, the attempt to cure genetic disease by addition of normally functioning genes from exogenous sources.

Genomics

The development of the technology to sequence the DNA of whole genomes on a routine basis has given rise to the discipline of genomics, which dominates genetics research today. Genomics is the study of the structure, function, and evolutionary comparison of whole genomes. Genomics has made it possible to study gene function at a broader level, revealing sets of genes that interact to impinge on some biological property of interest to the researcher. Bioinformatics is the computer-based discipline that deals with the analysis of such large sets of biological information, especially as it applies to genomic information.

Population Genetics

The study of genes in populations of animals, plants, and microbes provides information on past migrations, evolutionary relationships and extents of mixing among different varieties and species, and methods of adaptation to the environment. Statistical methods are used to analyze gene distributions and chromosomal variations in populations.

Population genetics is based on the mathematics of the frequencies of alleles and of genetic types in populations. For example, the Hardy-Weinberg formula, $p^2 + 2pq + q^2 = 1$, predicts the frequency of individuals with the respective homozygous dominant (*AA*), heterozygous (*Aa*) and homozygous recessive (*aa*) genotypes in a randomly mating population. Selection, mutation, and random changes can be incorporated into such mathematical models to explain and predict the course of evolutionary change at the population level. These methods can be used on alleles of known phenotypic effect, such as the recessive allele for albinism, or on DNA segments of any type of known or unknown function.

Human population geneticists have traced the origins and migration and invasion routes of modern humans, *Homo sapiens*. DNA comparisons between the present peoples on the planet have pointed to an African origin of *Homo sapiens*. Tracing specific forms of genes has allowed geneticists to deduce probable migration routes out of Africa to the areas colonized today. Similar studies show to what degree present populations have been mixed by recent patterns of travel.

Behavior Genetics

Another aspect of genetics is the study of the influence of heredity on behavior. Many aspects of animal behavior are genetically determined and can therefore be treated as similar to other biological properties. This is the subject material of behavior genetics, whose goal is to determine which genes control various aspects of behavior in animals. Human behavior is difficult to analyze because of the powerful effects of environmental factors, such as culture. Few cases of genetic determination of complex human behavior are known. Genomics studies provide a useful way to explore the genetic factors involved in complex human traits such as behavior.

Human Genetics

Some geneticists specialize in the hereditary processes of human genetics. Most of the emphasis is on understanding and treating genetic disease and genetically influenced ill health, areas collectively known as medical genetics. One broad area of activity is laboratory research dealing with the mechanisms of human gene function and malfunction and investigating pharmaceutical and other types of treatments. Since there is a high degree of evolutionary conservation between organisms, research on model organisms—such as bacteria, fungi, and fruit flies (*Drosophila*)—which are easier to study, often provides important insights into human gene function.

Many single-gene diseases, caused by mutant alleles of a single gene, have been discovered. Two well-characterized single-gene diseases include phenylketonuria (PKU) and Tay-Sachs disease. Other diseases, such as heart disease, schizophrenia, and depression, are thought to have more complex heredity components that involve a number of different genes. These diseases are the focus of a great deal of research that is being carried out today.

Another broad area of activity is clinical genetics, which centres on advising parents of the likelihood of their children being affected by genetic disease caused by mutant genes and abnormal chromosome structure and number. Such genetic counseling is based on examining individual and family medical records and on diagnostic procedures that can detect unexpressed, abnormal forms of genes. Counseling is carried out by physicians with a particular interest in this area or by specially trained non-physicians.

Molecular Basis of Inheritance

We are aware of the concepts of Inheritance, its pattern and genetic basis. As we know that the genetic basis resides in nucleic acid found in living beings. There are two types of nucleic acids:

- Deoxyribonucleic Acid (DNA);
- Ribonucleic Acid (RNA).

In living organisms, DNA act as the genetic material and RNA act as genetic material in some viruses, and play the role of messenger. DNA is located in nucleus primarily, and is also found in cell structures called mitochondria.

DNA

DNA was first isolated by Friedrich Miescher in the year 1869 and Francis Crick and James Watson identified its molecular structure in 1953.

Deoxyribonucleic acid is a molecule that carries the genetic instructions used in growth, development, functioning and reproduction of all known Living Organisms and many viruses. It is a long polymer of deoxyribonucleotides and length of DNA depends on the number of nucleotides present in it. Most DNA molecule has two biopolymer strands coiled around each other to form double helix. These two DNA strands are termed as polynucleotides as they are composed of simple monomer units is nucleotides.

Each nucleotide is composed of one of 4 nitrogen – containing nuclebases (such as cytosine (C), adenine (A), guanine (G) and thymine (T)), a phosphate group and a sugar called deoxyribose. These nucleotides are connected via covalent bond forming a chain between sugar of one nucleotide and the phosphate of the next, thereby resulting in an alternating sugar – phosphate backbone.

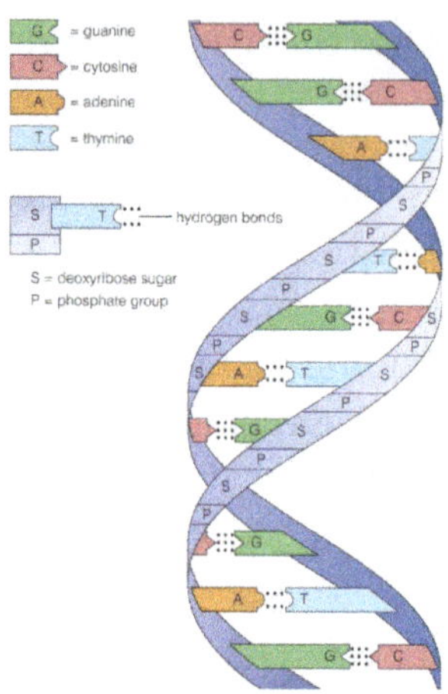

Figure shows the double helical structure of DNA with sugar – phosphate Backbone. DNA consists of biological information and its backbone are resistant to cleavage. Both the strands of DNA store same biological information and this information gets replicated when the two strands get separated. These two strands run in opposite direction and are antiparallel to each other.

Salient Features of Double Helical Structure of DNA

- DNA is composed of 2 polynucleotide chains and its backbone constitutes of sugar – phosphate and the bases project inside.
- Both the strands of DNA possess anti – parallel polarity. It implies, if the polarity of one strand is 5'à3' then the other strand will be 3'à5'.
- The bases in strands are paired via hydrogen bond (H – bond). It is important to note that Adenine forms two hydrogen bonds with Thymine from opposite strand and vice – versa. In the same way, Guanine bonds with cytosine with 3H – bond. As a result, always purine comes opposite to a pyrimidine.
- The two strands or chain are coiled in right – handed fashion and the pitch of helix is 3.4nm and there are around 10 bp in each turn. Thus, the distance between a bp in a helix is around 0.34 nm.
- The plane of one base pair stacks over another in double helix and this in addition to H – bond results in the stability of the helical structure.
- The double helical structure has major and minor grooves along the phosphodiester backbone.
- The genetic information resides on one of the two strands called sense strand or template strand.
- H – Bond is formed between purine and pyrimidine only. The one base arrangement possible in the structure of DNA are A – T, T – A, G – C and C – G.

Following figure shows the double stranded polynucleotide chain with all the features as discussed above, that is. 2 polynucleotide chains, its backbone with sugar – phosphate bond and its base arrangement (A – T, T – A, G – C and C – G):

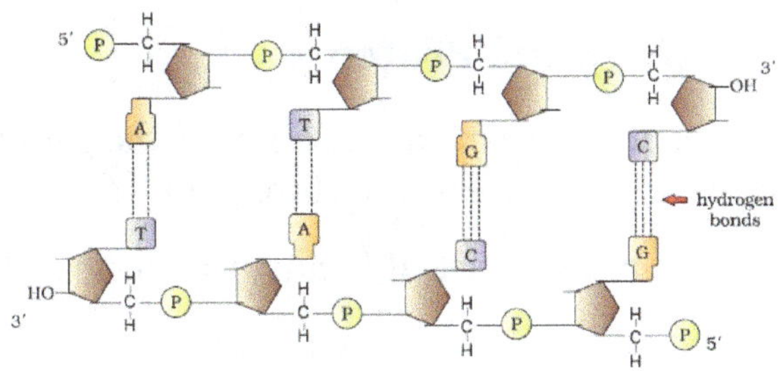

Each chromosome consists of one continuous thread - like molecule of DNA coiled tightly around proteins, and contains a portion of the 6,400,000,000 base pairs (DNA building blocks) that make up your DNA. The manner in which DNA is packaged into chromatin is factor that control the production of protein.

Packaging of DNA Helix

- Packaging in Prokaryotes: In prokaryotes well – defined nucleus is not present absent and therefore, DNA is held together with some proteins in a region called nucleoid. Here, the DNA is organized in the form of large loops held by proteins.
- Packaging in Eukaryotes: The organization of DNA is complex in eukaryotes. In the year 1974, Reger Kornberg reported that chromosome is composed of protein and DNA. On the basis of presence of amino acid residue with charged side chains, a protein gains its charge. Histones are the proteins that are rich in arginine and lysine residues. Both these amino acids carry positively charged and organize themselves to make a unit of 8 molecules referred as Histone Octamer. The negatively charged DNA is wrapped around positively charged Histone Octamer to form the structure called nucleosome.

Following figure shows the structure of nucleosome. It is made up of four types of proteins – H_2A, H_2B, H_3 and H_4 occurring in pairs:

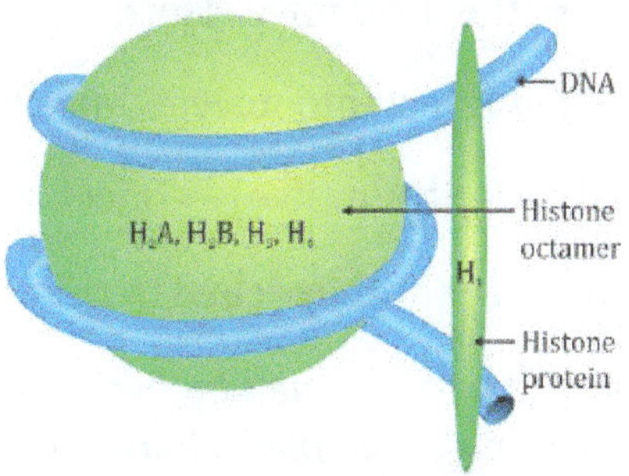

Structure of Nucleosome

One nucleosome comprise of 200 base pair of DNA helix and these are repeating units of chromatin which are thread like colored bodies in nucleus. When observed under electron microscope, it appears as "beads on string."

Non – Histone Chromosomal Proteins

These are additional set of proteins which contribute to the packaging of chromatin at higher level. Following regions are present in chromatin –

- Euchromatin – It is loosely packed in a typical nucleus. It is transcriptionally active.
- Heterochromatin – These are densely packed and are stains dark. It is transcriptionally inactive.

Search of Genetic Material

Transforming Principle – It was an early name for DNA. In the year 1928, scientists were not aware that DNA carried genetic information but they were aware that there was something that causes bacteria to transform from one form to another.

Frederick Griffith in the year 1928 carried out an experiment on pneumococcus bacteria. These bacteria were of two types, that is smooth type and rough type. When Streptococcus pneumoniae bacteria were grown on a culture plate, some produced rough colonies while some produced smooth shiny colonies. This is so because the S strain bacteria consist of polysaccharide (mucous) coat, while R was not having any coat. When a mouse was infected with S strain, they died due to pneumonia infection but mice infected with R strain do not developed any infection Thus,

S Strain → Inject into mice → Mice die

R Strain → Inject into mice → Mice live

Further, Griffith killed bacteria by heating them. He found that when heat – killed S Strain bacteria was injected, mice was alive (image below). Further, after injecting a mixture of heat – killed S and live R bacteria, the mice died. Added to this, he was able to recover living S bacteria from the dead mice. Thus,

S Strain (Heat – killed) → Inject into mice → Mice live

S Strain (Heat + killed) + R Strain (live) → Inject into mice → Mice die

Thus, from the above experiment, Griffith reached to the conclusion that –

"R Strain bacteria had somehow transformed by heat – killed S strain bacteria. This was guided by some "transforming principle." However, till this experiment, the biochemical nature of genetic material was not defined."

Following figure pictorially represents the entire experiment of Griffith. The above stated four reactions display the various conditions of mice in different stages:

R Strain → Inject into mice → Mice live

S Strain → Inject into mice → Mice die

S Strain (Heat – killed) → Inject into mice → Mice live

S Strain (Heat + killed) + R Strain (live) → Inject into mice → Mice die

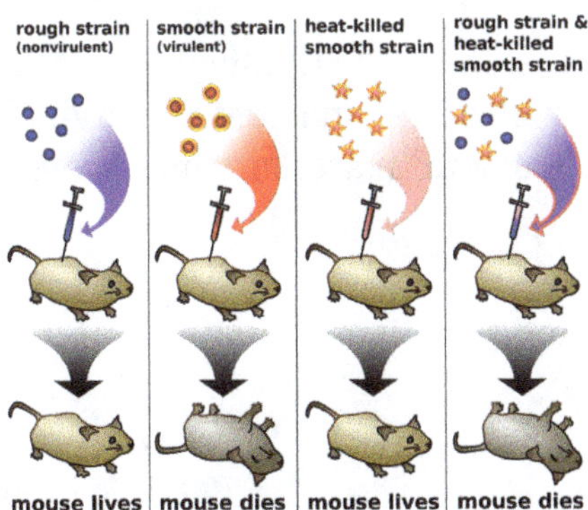

Biochemical Characterization of Transforming Principle

Initially, protein was considered as genetic material. In the year 1933 – 44, Oswald Avery, Colin Macleod and Maclyn McCarty worked on the objective of determining the biochemical nature of "transforming principle." So, they purified bio-chemicals like DNA, RNA, proteins, etc. from S cells in order to know which one can transform live R cells to S cells. They found:

- DNA from S bacteria is enough for transforming R bacteria.
- Proteases (protein digesting enzymes) and RNases (RNA digesting enzymes) have no effect on transformation. Thus, transforming substance was not RNA or protein.

Thus, they concluded that DNA was the only genetic material.

Genetic Material is DNA

Alfred Hershey and Martha Chase in 1952 studied bacteriophage that proved that *"DNA is the genetic material."* Bacteriophages are the viruses that can infect bacteria. These bacteriophages attaches to bacteria and its genetic material, thereby entering the bacterial cell. These cells treat viral genetic material and manufacture more virus particles. Hershey and Chase wanted to know whether it was DNA or protein from viruses that entered bacteria. Therefore, they grew some viruses in a special medium with radioactive phosphorus and some in the medium with radioactive sulfur.

Those radioactive phages were then allowed to associate to *E.Coli* bacteria and as the infection increased, the viral coats were eliminated from bacteria agitating them in blender. Thereby, separation of virus particles was done by spinning them in a centrifuge.

Those, bacteria which were infected by virus with radioactive DNA were radioactive, that helped in concluding that DNA is the material that was transferred from virus to bacteria. Added to this, the bacteria infected by radioactive protein were not radioactive. Thereby, this concluded that protein in not transferred from virus to bacteria.

Following figure shows all the steps of the Hershey – Chase experiments, which include infection at first, followed by blending and centrifugation. This experiment helped in concluding that DNA is the genetic material while protein in not the material that passed:

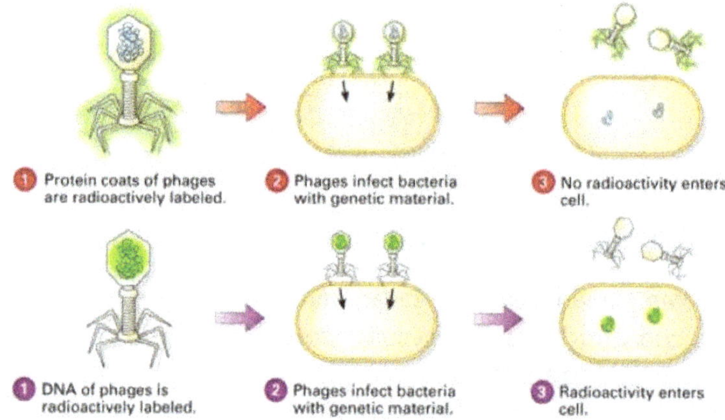

RNA World

RNA was the first genetic material. There are several evidences that suggest that essential life processes are evolved around RNA. RNA used to act as catalyst and genetic material. But RNA being a catalyst was reactive and therefore, unstable. This resulted in the evolution of DNA which is more stable. DNA is double stranded and resists change by evolving the process of repair.

DNA Replication

DNA replication is the process by which DNA makes a copy of itself during cell division.

Steps of DNA Replication

- At the first step, DNA 'unzips' its double helical structure. This is carried out by an enzyme called *helicase*. This helps in breaking of hydrogen bonds that hold commentary bases of DNA.

- The separation of two strands creates 'Y' shape called a replication fork. These two separated strands act as a template for making the new strands of DNA.
- One of strand is oriented in 3' to 5' direction which is referred as leading strand, while other strand is oriented in 5' to 3' direction referred as lagging strand. Due to this different orientation, the two strands replicated differently.

Following figure shows the replication of leading and lagging strands of DNA as discussed:

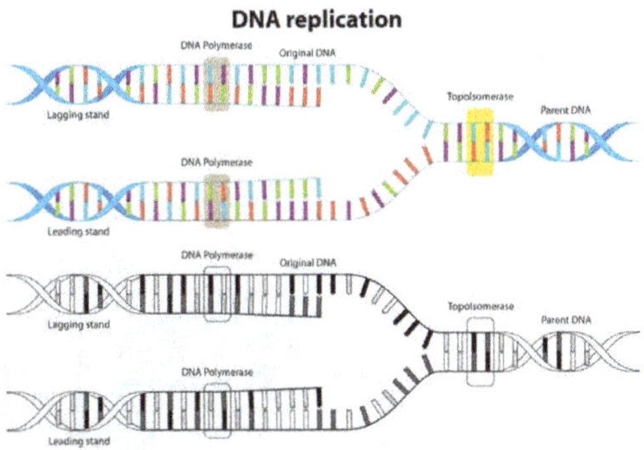

Leading strand	Lagging strand
• A short piece of RNA referred as primer comes along and binds to the end of the leading strand. This primer acts as starting point for the synthesis of DNA.	• Many primers are made from primase enzyme which binds at various points along the lagging strand.
• DNA polymerase binds to the leading strand and helps in adding the new complementary nucleotide bases to the strand of DNA in the direction 5' to 3'.	• Okazaki fragments (chunks of DNA) are then added to lagging strand in 5' to 3' direction.
• This is continuous replication.	• This is not continuous replication because Okazaki fragments will need to be joined later.

- After matching of all the bases, i.e. A - T and C – G, enzyme called exonuclease strips away the primer and those gaps were filled by complementary nucleotides.
- The newly formed strand is proofread in order to make sure that there is no mistake in the sequencing of DNA.
- In the end, DNA ligase, seals up the sequence of DNA into two continuous double strands. The result of this DNA replication is generation of two DNA molecules consisting of one new and one old chain of nucleotide. This scheme is referred as semiconservative DNA replication.

Transcription

The process of copying genetic information from one strand of the DNA into RNA is termed as transcription.

The process of transcription is governed by the principle of complementarity, except adenosine which forms base pair with uracil rather than thymine. Unlike the process of replication, in which total DNA is duplicated, in case of transcription, only a segment of DNA and only one strand is copied into RNA.

Transcription Unit

The transcription unit in DNA is defined primarily by three regions in DNA:

- A promoter
- The structural gene
- A terminator

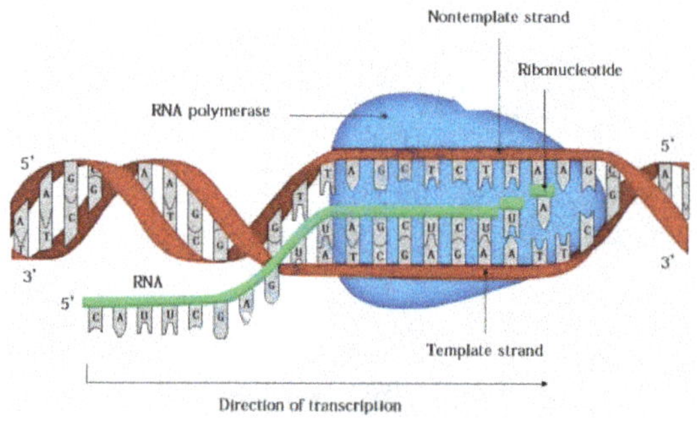

Transcription process

It is the first step of gene expression. At this step, the information from a gene is used to form a functional product called protein. The main objective of transcription is to make the copy of RNA of gene's DNA sequence. In case of protein coding gene, transcript or RNA copy carry the information required to form polypeptide.

Following figure shows the process of transcription and translation. In former, RNA makes the copy of DNA sequencing and this transcript carries the information required to form a polypeptide:

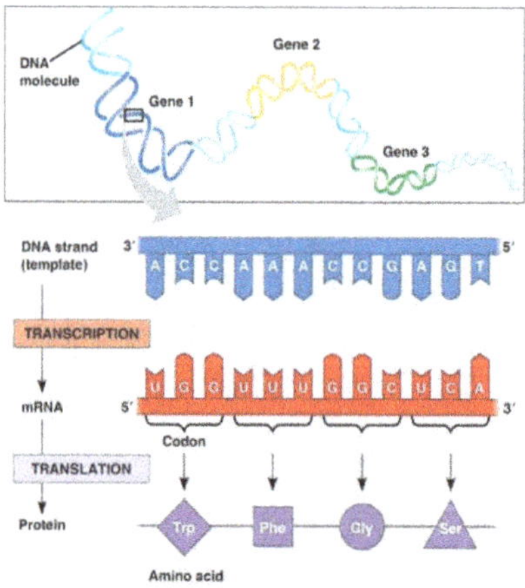

RNA Polymerase

It is the main enzyme involved in the process of transcription. It uses single stranded DNA template to form complementary strand of RNA. To be specific, RNA polymerase form RNA strand in 5' to 3' direction, by adding each new nucleotide to the 3' end of the strand.

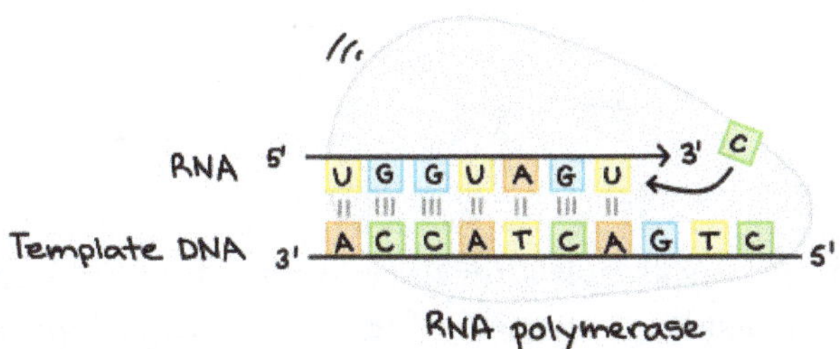

Stages of Transcription

Transcription of a gene is carried out in three stages, i.e. initiation, elongation and termination.

- Initiation: RNA polymerase binds itself to the sequence of DNA *(called promoter)* that is found near the beginning of a gene. Each gene has its own promoter and once bound, DNA strands gets separated from RNA polymerase providing the single stranded template required for transcription.

Following diagram shows the initiation stage in detail whereby, we can see the formation of single – stranded template.

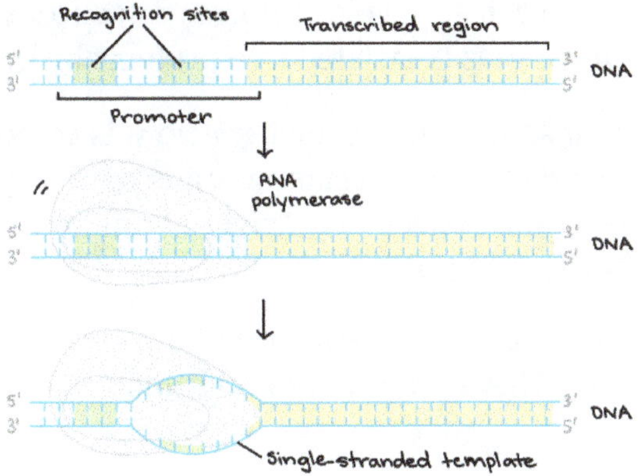

- Elongation: The template strand act as a template for RNA polymerase and as it refer this template, the polymerase build a RNA molecule out of complementary nucleotides, forming a chain which grows from 5' to 3'. The same information is carried by RNA transcript in the form of non – template strand of DNA, with the bases Uracil (U) in spite of Thymine (T).

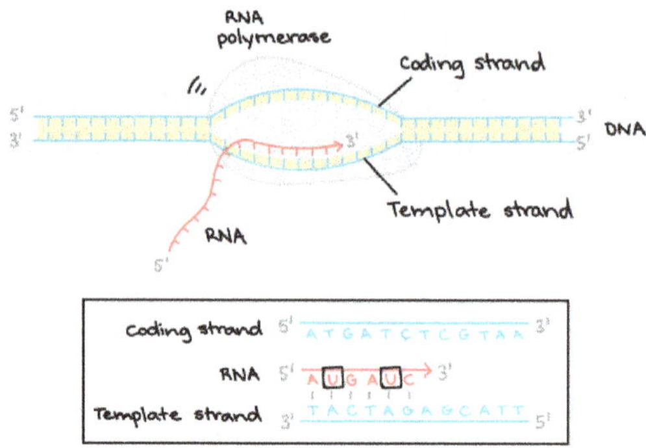

Elongation stage

Termination: Terminators signals after the completion of RNA and once the sequences are transcribed, they result in releasing of transcript from RNA polymerase.

Following diagram demonstrates the termination stage, which includes formation of a hairpin in the RNA:

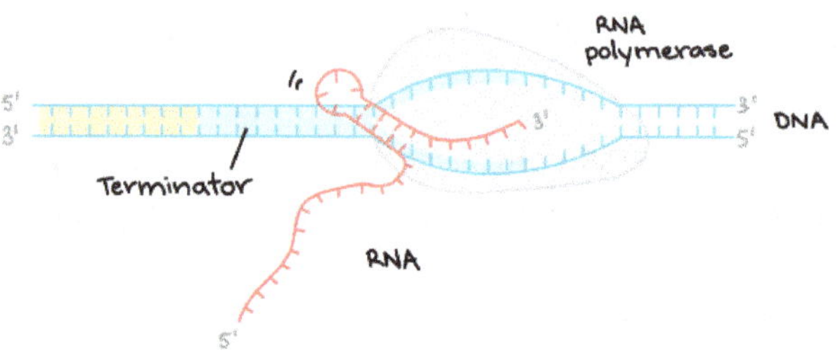

Genetic Code

The genetic code is the set of rules by which information encoded within genetic material (DNA or mRNA sequences) is translated into proteins by living cells.

During replication and transcription, one nucleic acid was copied forming another nucleic acid. On the other hand, during translation, transfer of genetic information is carried out from a polymer of nucleotides to polymer of amino acid. Thus, it can be said that genes are expressed in two steps:

- During transcription, the DNA sequence of a gene is rewritten in RNA and in case of eukaryotes, RNA goes through additional processing steps to become m-RNA.
- During translation, the sequence of nucleotides is translated into sequence of amino acids in protein chain.

Codon

Cells decode mRNAs by reading their nucleotides in groups of three, called codons.

Example of Codon

AUG, ATG, etc.

Several features of codon are as follows:

- Most codons specify amino acids.
- 3 “STOP” codon mark the end of protein.
- 1 “START” codon, AUG, is the beginning of protein and amino acid methionine is encoded.
- One codon codes form only 1 amino acid and therefore, it is specific in nature.
- Some amino acids are coded via more than one codon and thus, code is degenerate.

During translation, codon in mRNA is read, beginning with start codon, and continuing until stop codon is reached. These codons are read from 5’ to 3’ and specify the order of amino acid in proteins from N – terminus to C – terminus.

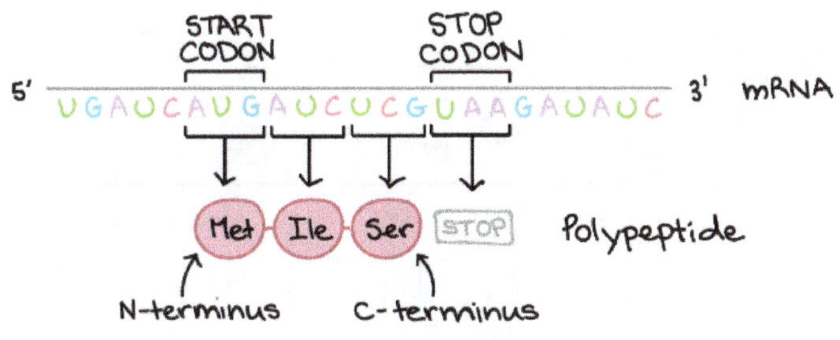

Working of codons

The full set of relationships between codons and amino acids (or stop signals) is called the Genetic Code.

Following table summarizes the genetic code. We can observe that many amino acids are represented in the table by more than one codon. For example, we can see there are six different methods to write Leucine in the language of mRNA:

Second letter

First letter	U	C	A	G	Third letter
U	UUU, UUC – Phe UUA, UUG – Leu	UCU, UCC, UCA, UCG – Ser	UAU, UAC – Tyr **UAA Stop** **UAG Stop**	UGU, UGC – Cys **UGA Stop** UGG Trp	U C A G
C	CUU, CUC, CUA, CUG – Leu	CCU, CCC, CCA, CCG – Pro	CAU, CAC – His CAA, CAG – Gln	CGU, CGC, CGA, CGG – Arg	U C A G
A	AUU, AUC, AUA – Ile AUG Met	ACU, ACC, ACA, ACG – Thr	AAU, AAC – Asn AAA, AAG – Lys	AGU, AGC – Ser AGA, AGG – Arg	U C A G
G	GUU, GUC, GUA, GUG – Val	GCU, GCC, GCA, GCG – Ala	GAU, GAC – Asp GAA, GAG – Glu	GGU, GGC, GGA, GGG – Gly	U C A G

Reading Frame

Reading frame helps in determining the way by which mRNA sequence is divided into codons during translation.

Example of Reading Frame

In the below figure, mRNA can encode three totally different proteins, on the basis of frame. Three different frames can be produced and start of codon's position ensures which frame is to be chosen:

Here the start codon is the key signal because translation starts at this codon and continues successively in the group of three.

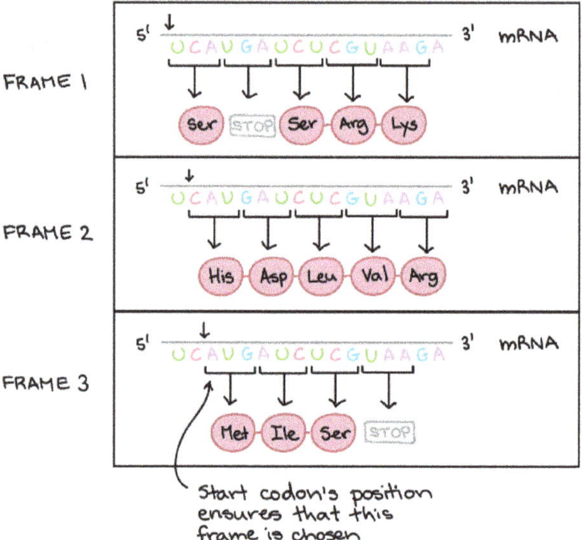

Translation

Translation refers to the process of polymerization of amino acids to form a polypeptide.

The order and sequence of amino acids are defined by the sequence of bases in mRNA and these amino acids are joined via peptide bond. The peptide bond is formed by the usage of energy and therefore, in the first step amino acids are activated and linked to cognate tRNA (referred as charging of tRNA or aminoacylation of tRNA). When these charged tRNAs are brought close, peptide bond is formed and this formation is catalyzed that enhances the rate of peptide bond formation.

References

- Overlapping-Genes-2634: medicine.jrank.org, Retrieved 19 April 2018

- Genetics, gene-types-and-functions-of-gene-12125: yourarticlelibrary.com, Retrieved 31 March 2018
- Introduction-to-genetics, biology-101: basicbiology.net, Retrieved 11 June 2018
- Genetics, science: britannica.com, Retrieved 25 May 2018
- Molecular-basis-of-inheritance, genetics-and-evolution, biology: emedicalprep.com, Retrieved 18 April 2018

Chapter 2

Genome

The genetic material of an organism is known as the genome. It is made up of DNA. The genes, mitochondrial DNA, noncoding DNA and chloroplast DNA together make up the genome. This chapter has been carefully written to provide an easy understanding of the basic concepts of genome size and complexity and coding regions. It also explores in depth the genome in the case of eukaryotic and prokaryotic organisms as well as in viruses.

A genome is the complete set of genetic information in an organism. It provides all of the information the organism requires to function. In living organisms, the genome is stored in long molecules of DNA called chromosomes. Small sections of DNA, called genes, code for the RNA and protein molecules required by the organism. In eukaryotes, each cell's genome is contained within a membrane-bound structure called the nucleus. Prokaryotes, which contain no inner membranes, store their genome in a region of the cytoplasm called the nucleoid. The full range of RNA molecules expressed by a genome is known as its transcriptome and the full assortment of proteins produced by the genome is called its proteome.

There are 23 pairs of chromosomes in the human genome. Between 1990 and 2003, all twenty-three pairs were fully sequenced through an international research undertaking known as the Human Genome Project. The study and analysis of genomes is called genomics.

Genomics main task is to determine the entire sequence of DNA or the composition of the atoms that make up the DNA and the chemical bonds between the DNA atoms. Knowledge of the DNA sequence has become an important part of biological research but it is also of vital importance in other research disciplines including medicine, biotechnology, forensic etc.

Genomics should not be confused with genetics. It is a study of the functions of single genes which shows a great potential in medicine and molecular biology as well. The field of genomics is interested in genome as a whole structure and investigates a single genome only if it is important for the genome as a structure. Genomics can therefore also be defined as a study of the complete genetic material of an organism.

History of genomics dates back to the 1970s when the scientists determined the DNA sequence of simple organisms. The greatest breakthrough in the field of genomics occurred in the mid-1990s when the scientists sequenced the entire genome of Haemophilus influenzae, a free-living organism which, however, does not cause influenza. The bacterium was thought to be the cause of flu until 1933 when it was proven that influenza is caused by a virus. In 2001, the scientists sequenced most of the human genome. Since then, genomes are being sequenced with relative ease. By the end of 2011, scientists sequenced genomes of over 2,700 viruses, more than 1,200 bacteria and archaea and 36 eukaryotes about 50 percent of which are fungi.

Scientists get a number of highly useful information from sequenced DNA of organisms. But what is most important of all, they allow the scientists to determine the relationships between the genes and different sections of DNA which in turn allows them to determine which areas could offer benefits to science as well as make the knowledge useful for medical applications.

Genomic research projects over the last few decades gave rise to several research areas in the study of genomes.

The main genomics research areas include:

Human genomics: Like its name suggests, human genomics is focused on studying the human genome sequence. Human DNA was sequenced by the Human Genome Project, an international scientific research project in 2001 but the human genome sequence was proclaimed completed only in 2007.

Bacteriophage genomics: It refers to the study of bacteriophage genomes or genomics of viruses which infect bacteria and are considered as a possible alternative for treatment of illnesses that are caused by antibiotic-resistant bacteria.

Metagenomics: It is a study of metagenomes or genetic material which is obtained from environmental samples rather than from cultivated cultures. Metagenomics has revolutionized the understanding of microbial world and shown that the traditional cultivation techniques have missed the majority of microbial diversity.

Cyanobacteria genomics: This field of genomic research is concentrated on study of cyanobacteria, a phylum of bacteria which get energy through photosynthesis.

Pharmacogenomics: This branch of genomics studies the impact of genetic variation on a drug's efficacy and toxicity, and plays an important role in optimization of drug therapy.

Viral Genomes

The genomes of viruses are just like every other genome: they contain all the genes necessary to replicate themselves. However, viruses are very lazy, and they use most of the host cell's machinery to make what they need to replicate. Kind of like that time we let that carnie into our house, then when we came back the dryer was gone and there were twelve carnies and a ring toss game in its place. Yeah, just like that.

Viruses have genes that encode a minimum of two proteins:

- Replicase – an enzyme that replicates the genome.
- Capsid – a protein that protects the genome.

Viruses may have genes that encode for proteins that aid in infection, survival, or other useful viral skills.

- Protease – enzyme that processes viral proteins and allows assembly or maturation to infect other virus cells.
- Glycoprotein – (enveloped viruses only) allows virus to enter a cell, targets specific cell types for virus, and aids in virus assembly.
- Host shutoff proteins – virus proteins that shut off host activities so only virus genes get made.
- Anti-host defense proteins – the number and type of these different genes will vary, depending on the virus. These proteins prevent the host defense mechanism from stopping virus replication.

These genes are encoded on either RNA or DNA polymers, and are either single stranded or double stranded.

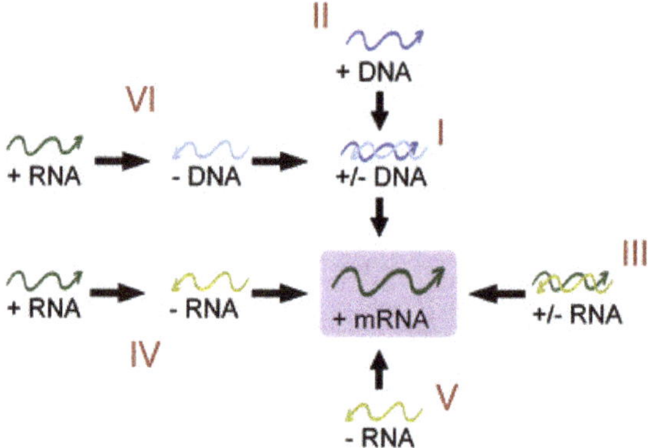

Morphological Features of Viruses

The viral chromosome is enclosed within a protein shell called capsid. The viral chromosome and its protein coat together are called nucleocapsid. Viruses vary considerably in their morphological features shown in table below:

Table: Capsid and nucleic acid characteristics of viruses

Morphological group	Virus	Nucleic acid
Helical capsid		
Naked	Coliphage fd	Single-stranded DNA

	Many plant viruses e.g.TMV. beet yellow	Single-stranded DNA
Enveloped	Influenza	Single-stranded segmented RNA
Icosahedral capsid		
Naked	Adeno associated	Single-stranded DNA
	Coliphage ϕ X174	Single-stranded DNA
	Polyoma	Double- stranded DNA
	Adenovirus	Double- stranded DNA
	Coliphage F_2	Single-stranded DNA
	Polio	Single-stranded DNA
	Many plant viruses e.g turnip yellow	Single-stranded DNA
	Reovirus	Double- stranded segment RNA
Enveloped*	Herpes simplex	Double- stranded DNA
Some components Helical and some Icosahedral (Naked)	Large bacterio-phages e.g T_2, T_4, T_6	Double- stranded DNA
Complex virions	Pox viruses, e.g Vaccinia, Variola. Viruses	Double- stranded DNA

* Enveloped Viruses are enclosed by a membrane of host origin.

1. Icosahedral virions

Their capsid is icosahedral, i.e. the virion is a regular polyhedron with 20 triangular faces and 12 corners. Examples are adenoviruses and bacteriophage φX174.

2. Helical virions

The nucleic acid of such virions is enclosed in a cylindrical, rod shape capsid that forms a helical structure, e.g., TMV, bacteriophage M13.

3. In some cases, the nucleocapsid is icosahedral while in others, it is helical in some components. Such viruses are enveloped.

4. Coplexvrions

These viruses do not have a clearly identifiable capsid. The viral nucleic acid is present in the centre of the shell which is made up of protein molecules. Some of the shells are complex while others are simple. In Herpes, an animal virus that contains DNA as genetic material, the capsid has a diameter of 1000A; it is further surrounded by an envelope making its diameter 1500A shown in figure below:

The capsid is mode up of protein subunits (capsomers) which form an icosahedron.

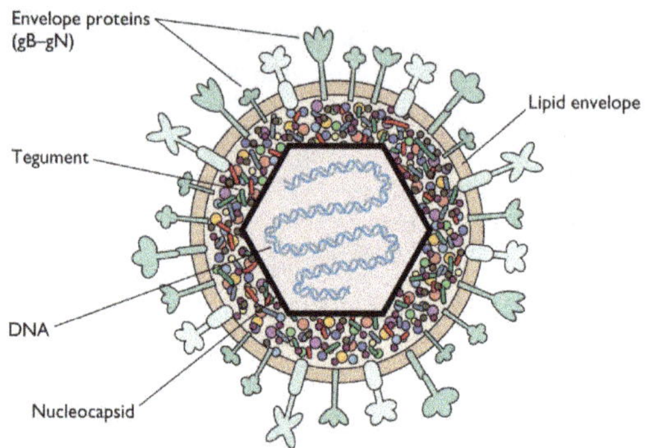

Schematic representation of the general structure of Herpes virus. It contains DNA and multiplies in animals cells. DNA containing core is embedded in the icosahedral capsid which is made up of protein molecules called capsomers. The capsid is surrounded by an envelope.

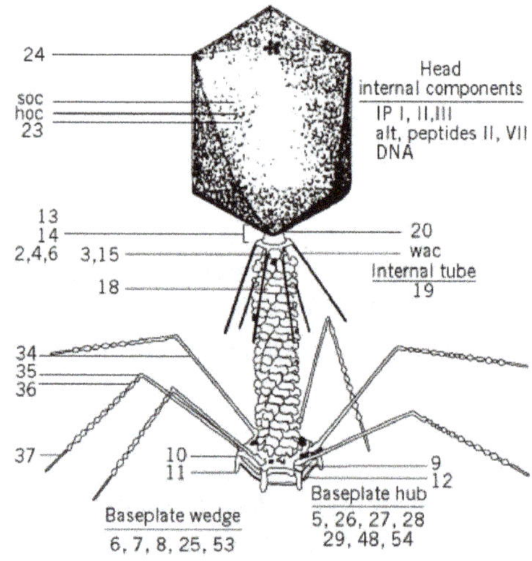

Structure of a T-even phage (T2, T4, T6). The head (capsid) contains DNA that is about 170 kb in length. Fibres bind to specific sites on the host cell. DNA is injected into the cell though the tube

Bacteriophages have relatively complex structures: they contain a head, a tail, a base plate and several tail fibres shown in just above figure. The head is hexagonal (lateral side) and contains the viral DNA. The tail has a core tube surrounded by a sheath. At the tail end, there is a basal plate with 6 spikes from which 6 tail fibres emerge.

At the time of infection, the tail fibres bind to specific receptor sites on the host cell. The base plate is drawn to the cell surface and contraction of tube sheath occurs along with the removal of the base plate plug. The core of the tail penetrates the cell wall which is weakened by some hydrolytic enzymes present in the phage and the viral tail. DNA enters into the host cell through the core tube of the tail.

In the case of tobacco mosaic virus (TMV; multiplying in tobacco plant cells) and some small bacterial viruses, the protein coat contains a single type of protein. These protein molecules are arranged in either a helical symmetry or a cubical symmetry.

The shell of TMV contains about 2150 protein molecules which are identical, each molecule having the molecular weight of-17,000. These molecules are helically arranged around the RNA genome which contains 6,000 nucleotides.

Lytic Viruses

The viruses which lyse or disrupt the host cell following infection are called lytic viruses. During infection, the nucleic acid is injected into the host cell. The enzymes required for viral DNA replication are then synthesized so that replication of DNA occurs to produce numerous copies of the viral chromosome.

The protein components of the capsid are synthesized in the later stages leading to the formation of heads and tails; the viral DNA is then packed into the heads. In the end, the cell wall ruptures and the progeny phage particles are released show in figure:

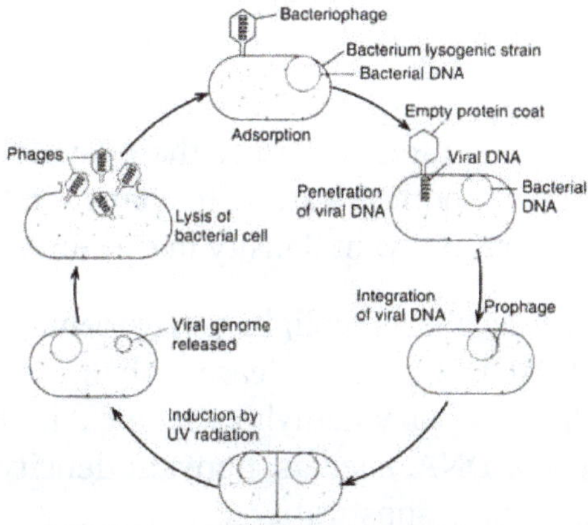

Fig. 2.46 : Lysogenic life-cycle of bacteriophage (λ-phage)

Life cycle of lytic bacteriophage

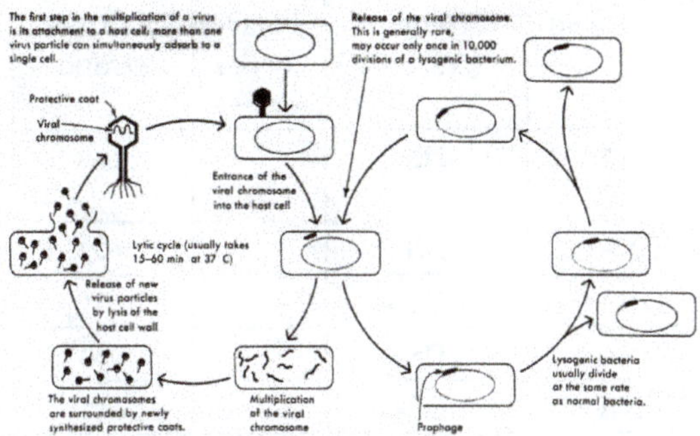

Life cycle of temperate phages

The virulent state of such phages showing their multiplication and the resultant lysis of their host cell: this releases new viral particles. B Viral chromosome integrates into the host chromosome (lysogenic state) and becomes prophage (provirus). After induction, the viral chromosome is excised from the host chromosome and becomes lytic.

Lysogenic Viruses (Temperate Phages)

Lysogeny involves a symbiotic relationship between a temperate phage and its bacterial host. The viral chromosome becomes inserted into the bacterial chromosome, where it remains and replicates along with the latter. The viral DNA integrated into the bacterial genome is called a provirus or prophage shown in just above figure. The bacterium containing a prophage is immune to the infection by the same virus.

Viral Chromosomes

Viruses contain either DNA or RNA as their genetic material. These nucleic acids may be either single or double-stranded. Small viruses may contain 3 kb (kb =,kilo-bases = 1000 bases), while large viruses could have about 300 kb. in their genome. Thus the number of genes in viral genome may vary from only 3 to hundreds. The retroviruses arc diploid (have two copies of the genome per capsid), while the others are haploid.

Double-stranded DNA

Several viruses possess double-stranded DNA as their genetic material. The base composition of different viruses is modified leading to change in the physical properties of DNA such melting temperature, buoyant density in caesium chloride (CsCl) etc.

In some of the viruses, such as T-even coliphages, cytosine (C) is modified into 5-hydroxymethyl- cytosine (HMC). In certain cases, thymine is converted into 5-hydroxy-methyl uracil or 5-di-hydroxymethyluracil, e.g., in B. subtilisbacteriophges. Certain physical properties of DNA, such as, buoyant density in CsCl or melting temperature are changed due to these substitutions.

Table: Viral nucleic acids and their characteristics

Virus	Nucleic acid	Strands* (DS/SS)	Polarity (+) or (−)	Kilobases/ strand	Mol. Wt. (10^6daltons)	Number of segmants
Vaccinia virus(a pox virus)	DNA	DS		242	160	
B. subtilis phage SP8	DNA	DS		197	130	
Coliphage T2, T6	DNA	Ds		167	110	
Coliphage T4	DNA	DS		170		
Herpes simplex	DNA	DS		151	100	
Coliphage T5	DNA	DS		117	77	
Coliphage λ	DNA	DS		47	31	
Coliphage T7	DNA	DS		38	25	
Adenovirus Type 2	DNA	DS		35	23	
Polyoma virus	DNA	DS (circular)		5	3.5	

Coliphage M 13	DNA	SS(circular)		7.3	2.4	
Coliphage ϕ 174	DNA	SS(circular)		5.2	1.7	
Phage fd	DNA	SS		6		
Rice dwarf virus	RNA	DS		23	15**	10
Reovirus	RNA	DS		23	15**	10
Influenza virus (a myxovirus)	RNA	SS	(-)	18	6**	8
Rous sarcoma virus (a retrovirus) #	RNA	SS	(+)	10.5	3.5	
Bunya virus	RNA	SS	(-)	9	3**	3
Polio. Virus (a Picorna virus)	RNA	SS	(+)	7.5	2.5	
Turnip yellow mosaic	RNA	SS	(+)	6	2	
Coliphage R 17	RNA	SS	(+)	4	1.3	
TMV	RNA	SS	(+)	6.4	2	

Some of the viruses contain linear DNA, while others contain circular (cyclic) DNA shown in table above. In the case of phage lambda (λ), DNA can exist in both linear and cyclic forms. When isolated from a viral particle, the λ DNA is linear, but when it enters into the host cell, becomes circular. However, it enters into the host cell in its linear form.

The A. chromosome is a double- stranded DNA molecule containing 47,000 nucleotides; it is 17 pm in length. There is single- stranded projection of 12 nucleotides at each 5′-end; these projections are complementary to each other and thus they are called cohesive ends.

These cohesive ends are responsible for the circularization of the chromosome. Circularization of the chromosome protects it from degradation by the host exonucleases. Further, the linear DNA cannot replicate vegetatively; the circularity therefore, provides an advantage in replication as well.

Single-stranded DNA

Single-stranded DNA occurs in very small bacteriophages shown in table. The single-stranded DNA found in the virion is called the positive (+) strand; as a rule only the plus (+) strand is found in the phage particles. However, in adeno-associated viruses, two complementary strands exist in different virions. The single-stranded DNA contains inverted repeating sequences that form hair pins. The hairpin structures have important role in circularization of the linear strands and in replication.

Double-stranded RNA

Double-stranded RNAs are found in several icosahedral viruses of animals and plants. The genomes of such viruses are segmented. The different segments may be connected short stretches of base pairs. Transcription of each segment occurs separately and the enzyme involved is "Double-stranded RNA transcriptase". Each mRNA, on translation produces a separate polypeptide chain.

Single-stranded RNA

Single-stranded RNA is the genetic material in a number of viruses. Some viruses contain a single RNA molecule in their genome, while some other viruses contain several segments, e.g., influenza virus has 8 segments. The viruses contain either positive (+) or negative (-) strands of RNA in their capsids.

Positive (+) Strand

The viral RNA strand that functions as mRNA in the host cell is called the plus (+) strand or positive strand. The RNA genomes of animal viruses have a cap at their 5′-end and a poly (A) sequence at the 3′-end. However, in Picornavirus RNA, there is a special sequence at the 5′-end to which a small protein is covalently attached.

The RNA genomes of plant viruses possess a cap at the 5′-end but they do not contain the poly (A) at their 3′-ends; their 3′-end is similar to tRNA. Each retrovirus particle contains two copies of the (+) RNA strand representing its genome; these copies are held together near the 5′-end.

Negative (-) Strand

These RNAs do not contain a cap but terminate into a nucleoside triphosphate at their 5′-ends. These strands do not function as mRNA directly. Instead, they are transcribed by the enzyme "single-stranded RNA transcriptase" present in the virion, to produce the mRNA.

Packaging of Nucleic Acids in the Viruses

Viral genome (DNA/RNA) is tightly packed into the protein shell (capsid). The density of the nucleic acid in the protein shell is higher than 500 mg/ml, which is much greater than the density of DNA in other organisms. For example, density of DNA in bacterium is about 10 mg/ml, while in the eukaryotic nucleus, it is about 100 mg/ml. This shows that the nucleic acid is very tightly packaged in the viral particles.

The genetic material of TMV is single-stranded RNA containing 6400 nucleotides, making up a length of 2 pm. This RNA is packaged into the rod-shaped compartment of 0.3 x 0.008 pm. Adenoviruses contain 11 pm long double-stranded DNA consisting of 35,000 bp: this is packaged into an icosahedron type capsid of 0.07 pm diameter.

Phage T4 has a very long double-stranded DNA molecule (55 pm) having 170,000 bp. The capsid containing this rather long DNA is an icosahedron with the dimensions of

1.0 x 0.065 pm. Unlike eukaryotic nucleus and bacterial nucleoid, the volume of the capsid is fully packaged with the nucleic acid.

Packaging of nucleic acid to form a nucleocapsid occurs in two general ways. In one mechanism, the protein molecules assemble around the nucleic acid, e.g., in TMV. In the other mechanism, the protein coat is formed first and then the nucleic acid is inserted in it. In TMV, a duplex hairpin structure occurs in the RNA.

The assembly of protein monomers begins at this nucleation centre and proceeds in both the directions, reaching the ends. A total of 17 protein units form a circular layer and two such layers together form a unit of capsid. This structure interacts with the RNA which is coiled to form a helix inside the shell.

In bacteriophage T4 and λ etc., the protein shell is formed first. The nucleic acid is inserted into the coat from one end and then the tail is joined to the head. In case of circular DNA, it must be first converted into a linear molecule for packaging.

The lambda (λ) genome is circular and contains two "cos" sites, cosL and cosR. The free end in λ DNA is produced by enzymatic cleavage at the cosL site. Insertion of DNA occurs from this end and continues till the cosR site enters the capsid; a cleavage then occurs at the cosR site to produce the other end of the λ genome.

Some of the viruses, e.g., phage T4 and λ. have terminal redundancy in their genomes. In these viruses, multiple genomes join end-to-end to produce "concatemeric structure." In case of T4, insertion of the viral chromosome starts at a "random" point and continues until the required amount of DNA has been inserted into the head. The DNA inserted into the head has a terminal redundancy.

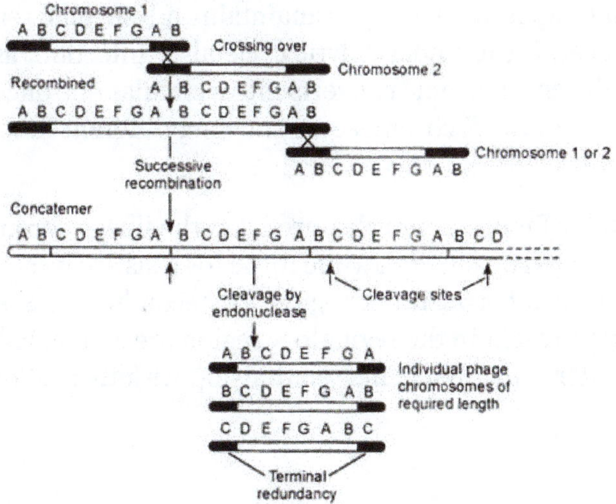

Formation of a tandem series of phage genomes (concatemers) through successive recombination between individual genomes of normal length. Specific endonuclease cuts the concatermeric DNA into required size to be packaged into the capsid. The terminal redundancy may lead to heterozygosity.

One likely origin of the "concatermeric" DNA is recombination. Recombination between two chromosomes combines two genomes end-to-end. Then recombination with a third genome produces a concatemer through successive recombination's.

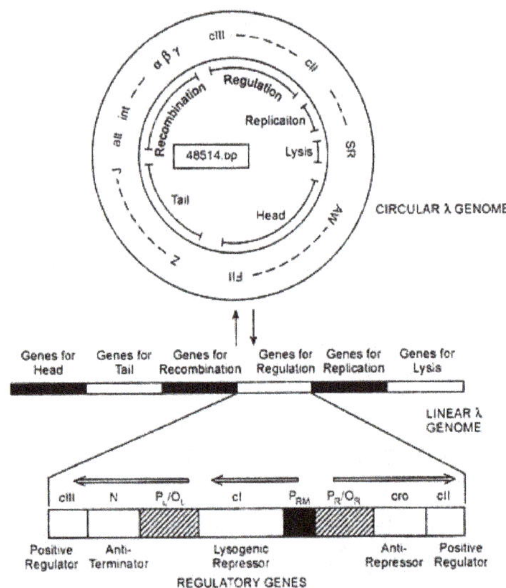

Another mechanism suggested for concatemer formation is the rolling circle replication. Specific endonuclease cuts the concatemer at the points that produce the genome of the "required length." The genomic DNA has homologous ends due to the terminal redundancy. Therefore, some chromosomes may be heterozygous for the terminal genes.

Mechanisms of Lysogenic and Lytic Pathways

Bacteriophage λ is a temperate phage that maintains a lysogenic relationship with its bacterial host. However, it can undergo lytic cycle also. Infection, as a rule, occurs in the linear form, but the chromosome converts into a circular one once it enters the host cell. A generalized map of the X chromosome showing different functions is presented in figure above.

Genes related to similar functions are clustered. On the linear chromosome, genes for head formation are located on left end, while those for lysis are located at the right end. The regulatory region lies between the region for recombination and the region for replication. The genes present in the regulatory region are responsible for determining whether the X will enter into a lysogenic relationship with its host or it will follow the lytic pathway.

Lytic Pathway

Regulatory genes are clustered and flanked by genes for recombination on their left side and those for replication on the right side (Fig. above). Genes N (anti-terminator) and era (anti-repressor) are located within the regulatory region. These genes are called "immediate early genes"; they are transcribed by the host RNA polymerase.

In the presence of anti-termination factor (p^N), transcription of both the genes (N and era) continues. These two genes are transcribed from different DNA strands in the opposite direction, the gene N being transcribed towards the left, while era is transcribed towards the right.

The transcription extends to other region of the genome for different functions (Fig. below). In the absence of cl repressor protein, the host RNA polymerase binds to P_L/O_L sites so that the transcription of the "late genes" is initiated; as a result, phage particles are produced and the cell is lysed.

Lysogenic Pathway

The regulatory region contains the cl gene which is responsible for the lysogenic pathway. A mutation in this region causes the phage to undergo lytic cycle.

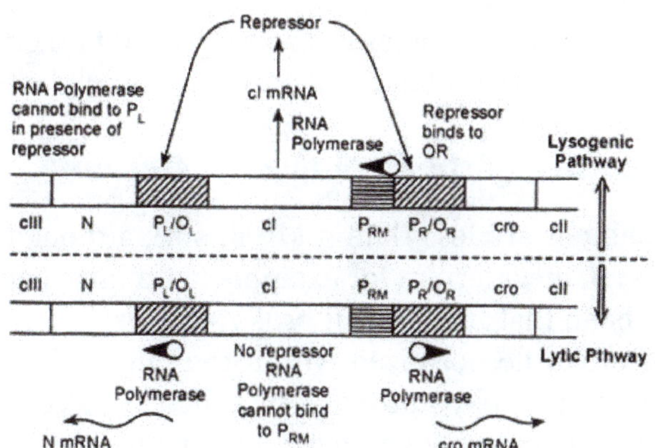

Mechanism of lytic and lysogenic pathways

The cl gene is transcribed to produce mRNA; the enzyme involved in transcription is RNA polymerase that binds to the promoter for repressor maintenance (P_{RM}). The transcription occurs from right to left. This cl mRNA is translated to produce the repressor monomer (Fig. above).

Repressor dimers are formed that bind to the P_L/O_R and P_L/O_L sites, thus preventing the RNA polymerase from binding to these promoters. This leads to the inhibition of transcription of N and cro genes. Later, the X chromosome is integrated into the bacterial chromosome; its delayed early genes are not expressed and the phage remains as a "provirus". Delayed early genes are the genes for recombination, replication and Q (anti-terminator). Late genes are tail, head and lysis genes.

When the cl repressor is bound to the o_L and o_R sites, RNA polymerase initiates transcription of the cl gene, and synthesis of repressor protein is continued. But in absence of the repressor, RNA polymerase binds to P_L/O_L and P_r/O_r sites and transcription of N and cro genes begins.

Thus the presence of cl repressor itself is necessary for its synthesis. Continuous production of cl repressor is necessary for lysogeny to be maintained. During this period the O_L and O_R sites are always bound by repressor.

When the lysogenized cell is infected by another phage X, the cl repressor protein produced by the "prophage" immediately binds to O_L and O_R sites of the infecting X genome. The function of the infecting X genes is thus inhibited and the cell remains immune to X infection.

Prokaryotic Genomes

The genomes of Bacteria and Archaea are compact; essentially all of their DNA is "functional" (contains genes or gene regulatory elements). Genes in a biochemical pathway or signaling pathway are often clustered together and arranged into operons, where they are transcribed as a single mRNA that is translated to make all the proteins in the operon. The size of prokaryotic genomes is directly related to their metabolic capabilities – the more genes, the more proteins and enzymes they make.

The Physical Structure of the Prokaryotic Genome

Most prokaryotic genomes are less than 5 Mb in size, although a few are substantially larger than this: *B.* megaterium for example has a huge genome of 30 Mb. The traditional view has been that in a typical prokaryote the genome is contained in a single circular DNA molecule, localized within the nucleoid – the lightly staining region of the otherwise featureless prokaryotic cell. This is certainly true for *E. coli*and many of the other commonly studied bacteria. However, our growing knowledge of prokaryotic genomes is leading us to question several of the preconceptions that became established during the pre-genome era of microbiology. These preconceptions relate both to the physical structure of the prokaryotic genome and its genetic organization.

The Traditional View of the Bacterial 'Chromosome'

As with eukaryotic chromosomes, a prokaryotic genome has to squeeze into a relatively tiny space (the circular *E. coli* chromosome has a circumference of 1.6 mm whereas an *E. coli* cell is just 1.0 × 2.0 μm) and, as with eukaryotes, this is achieved with the help of DNA-binding proteins that package the genome in an organized fashion. The resulting structure has no substantial similarities with a eukaryotic chromosome, but we still use 'bacterial chromosome' as a convenient term to describe it.

Most of what we know about the organization of DNA in the nucleoid comes from studies of *E. coli*. The first feature to be recognized was that the circular *E. coli* genome is supercoiled. Supercoiling occurs when additional turns are introduced into the DNA double helix (positive supercoiling) or if turns are removed (negative supercoiling). With a linear molecule, the torsional stress introduced by over- or under-winding is immediately released by rotation of the ends of the DNA molecule, but a circular molecule, having no ends, cannot reduce the strain in this way. Instead the circular molecule responds by winding around itself to form a more compact structure. Supercoiling is therefore an ideal way to package a circular molecule into a small space. Evidence that supercoiling is involved in packaging the circular *E. coli* genome was first obtained in the 1970s from examination of isolated nucleoids, and subsequently confirmed as a feature of DNA in living cells in 1981. In *E. coli,* the supercoiling is thought to be generated and controlled by two enzymes, DNA gyrase and DNA topoisomerase I.

Studies of isolated nucleoids and of living cells have shown that the *E. coli* DNA molecule does not have unlimited freedom to rotate once a break is introduced. The most likely explanation is that the bacterial DNA is attached to proteins that restrict its ability to relax, so that rotation at a break site results in loss of supercoiling from only a small segment of the molecule. The current model has the *E. coli* DNA attached to a protein core from which 40-50 supercoiled loops radiate out into the cell. Each loop contains approximately 100 kb of supercoiled DNA, the amount of DNA that becomes unwound after a single break.

The protein component of the nucleoid includes DNA gyrase and DNA topoisomerase I, the two enzymes that are primarily responsible for maintaining the supercoiled state, as well as a set of at least four proteins believed to have a more specific role in packaging the bacterial DNA. The most abundant of these packaging proteins is HU, which is structurally very different to eukaryotic histones but acts in a similar way, forming a tetramer around which approximately 60 bp of DNA becomes wound. There are some 60 000 HU proteins per *E. coli* cell, enough to cover about one-fifth of the DNA molecule, but it is not known if the tetramers are evenly spaced along the DNA or restricted to the core region of the nucleoid.

Complications on the E. Coli Theme

In recent years it has become clear that the straightforward view of prokaryotic genome anatomy developed from studies of *E. coli* is an over-simplification. Although the majority of bacterial and archaeal chromosomes are indeed circular, an increasing number of linear ones are being found. The first of these, for *Borrelia burgdorferi,* the organism that causes Lyme disease, was described in 1989 and during the following years similar discoveries were made for *Streptomyces* and other bacteria.

A second complication concerns the precise status of plasmids with regard to the prokaryotic genome. A plasmid is a small piece of DNA, often, but not always circular, that coexists with the main chromosome in a bacterial cell. Some types of plasmid are able to integrate into the main genome, but others are thought to be permanently independent. Plasmids carry genes that are not usually present in the main chromosome, but in many cases these genes are non-essential to the bacterium, coding for characteristics such as antibiotic resistance, which the bacterium does not need if the environmental conditions are amenable. As well as this apparent dispensability, many plasmids are able to transfer from one cell to another, and the same plasmids are sometimes found in bacteria that belong to different species. These various features of plasmids suggest that they are independent entities and that in most cases the plasmid content of a prokaryotic cell should not be included in the definition of its genome.

With a bacterium such as *E. coli* K12, which has a 4.6-Mb chromosome and can harbor various combinations of plasmids, none of which is more than a few kb in size and all of which are dispensable, it is acceptable to define the main chromosome as the 'genome'. With other prokaryotes it is not so easy. *Vibrio cholerae* the pathogenic bacterium that causes cholera, has two circular DNA molecules, one of 2.96 Mb and the other of 1.07 Mb, with 71% of the organism's 3885 genes on the larger of these. It would appear obvious that these two DNA molecules together constitute the *Vibrio* genome, but closer examination reveals that most of the genes for the central cellular activities such as genome expression and energy generation, as well as the genes that confer pathogenicity, are located on the larger molecule. The smaller molecule contains many essential genes but also has certain features that are considered characteristic of plasmids, notably an integron, a set of genes and other DNA sequences that enable plasmids to capture genes from bacteriophages and other plasmids. It therefore appears possible that the smaller genome is a 'megaplasmid' that was acquired by the ancestor to *Vibrio* at some period in the bacterium's evolutionary past. *Deinococcus radiodurans* R1, whose genome is of particular interest because it contains many genes that help this bacterium resist the harmful effects of radiation, is constructed on similar lines, with essential genes distributed among two circular chromosomes and two plasmids. However, the *Vibrio* and *Deinococcus* genomes are relatively non-complex compared with *Borrelia burgdorferi* B31, whose linear chromosome of 911 kb, carrying 853 genes, is accompanied by 17 or 18 linear and circular plasmids which together contribute another 533 kb and at least 430 genes. Although the functions of most of these genes are unknown, those that have been identified include several that would not normally be considered dispensable, such as genes for membrane proteins and purine biosynthesis. The implication is that at least some of the *Borrelia* plasmids are essential components of the genome, leading to the possibility that some prokaryotes have highly multipartite genomes, comprising a number of separate DNA molecules, more akin to what we see in the eukaryotic nucleus rather than the 'typical' prokaryotic arrangement. This interpretation of the *Borrelia* genome is still controversial, and is complicated by the fact that the related bacterium *Treponema pallidum*, whose genome is a single circular DNA molecule of 1138 kb containing 1041 genes, does not contain any of the genes present on the *Borrelia* plasmids.

The final complication regarding the physical structures of prokaryotic genomes concerns differences between the packaging systems for bacterial and archaeal DNA molecules. One reason why the archaea are looked upon as a distinct group of organisms, different from the bacteria, is that archaea do not possess packaging proteins such as HU but instead have proteins that are much more similar to histones. Currently we have no information on the structure of the archaeal nucleoid, but the assumption is that these histone-like proteins play a central role in DNA packaging.

The Genetic Organization of the Prokaryotic Genome

We have already learnt that bacterial genomes have compact genetic organizations with very little space between genes. To re-emphasize this point, the complete circular gene map of the *E. coli* K12 genome. There *is* non-coding DNA in the *E. coli* genome, but it accounts for only 11% of the total and it is distributed around the genome in small segments that do not show up when the map is drawn at this scale. In this regard, *E. coli* is typical of all prokaryotes whose genomes have so far been sequenced – prokaryotic genomes have very little wasted space. There are theories that this compact organization is beneficial to prokaryotes, for example by enabling the genome to be replicated relatively quickly, but these ideas have never been supported by hard experimental evidence.

Operons are Characteristic Features of Prokaryotic Genomes

One characteristic feature of prokaryotic genomes illustrated by *E. coli* is the presence of operons. In the years before genome sequences, it was thought that we understood operons very well; now we are not so sure.

An operon is a group of genes that are located adjacent to one another in the genome, with perhaps just one or two nucleotides between the end of one gene and the start of the next. All the genes in an operon are expressed as a single unit. This type of arrangement is common in prokaryotic genomes. A typical *E. coli* example is the lactose operon, the first operon to be discovered, which contains three genes involved in conversion of the disaccharide sugar lactose into its monosaccharide units – glucose and galactose. The mono saccharides are substrates for the energy-generating glycolytic pathway, so the function of the genes in the lactose operon is to convert lactose into a form that can be utilized by *E. coli* as an energy source. Lactose is not a common component of *E. coli*'s natural environment, so most of the time the operon is not expressed and the enzymes for lactose utilization are not made by the bacterium. When lactose becomes available, it switches on the operon; all three genes are expressed together, resulting in coordinated synthesis of the lactose-utilizing enzymes. This is the classic example of gene regulation in bacteria.

Altogether there are almost 600 operons in the *E. coli* K12 genome, each containing two or more genes, and a similar number are present in *Bacillus subtilis*. In most cases the genes in an operon are functionally related, coding for a set of proteins that are involved in a single biochemical activity such as utilization of a sugar as an energy source or synthesis of an amino acid. An example of the latter is the tryptophan operon of *E. coli*. Microbial geneticists are attracted to the simplicity of this system whereby a bacterium is able to control its various biochemical activities by regulating the expression of groups of related genes linked together in operons. This may be a correct interpretation of the function of operons in *E. coli*, *Bacillus subtilis* and many other prokaryotes, but in at least some species the picture is less straightforward. Both the archaeon *Methanococcus jannaschii* and the bacterium *Aquifex aeolicus* have operons, but the genes in an individual operon rarely have any biochemical relationship. For example, one of the operons in the *A. aeolicus* genome contains six linked genes, these genes coding for two proteins involved in DNA recombination, an enzyme used in protein synthesis, a protein required for motility, an enzyme involved in nucleotide synthesis, and an enzyme for lipid synthesis. This is typical of the operon structure in the *A.* aeolicus and *M. jannaschii* genomes. In other words, the notion that expression of an operon leads to the coordinated synthesis of enzymes required for a single biochemical pathway does not hold for these species.

Genome projects have therefore confused our understanding of operons. It is certainly too early to abandon the belief that operons play a central role in biochemical regulation in many bacteria, but we need to explain the unexpected features of the operons in *A.*aeolicus and *M. jannaschii*. It has been pointed out that both *A.* aeolicus and *M.* jannaschii are autotrophs, which means that, unlike many prokaryotes, they are able to synthesize organic compounds from carbon dioxide, but how this similarity between the species might be used to explain their operon structures is not clear.

Prokaryotic Genomes and the Species Concept

Genome projects have also confused our understanding of what constitutes a 'species' in the prokaryotic world. This has always been a problem in microbiology because the standard biological definitions of species have been difficult to apply to microorganisms. The early taxonomists such as Linneaus described species in morphological terms, all members of one species having the same or very similar structural features. This form of classification was in vogue until the early 20th century and was first applied to microorganisms in the 1880s by Robert Koch and others, who used staining and biochemical tests to distinguish between bacterial species. However, it was recognized that this type of classification was imprecise because many of the resulting species were made up of a variety of types with quite different properties. An example is provided by *E. coli* which, like many bacterial species, includes strains with distinctive pathogenic characteristics, ranging from harmless through to lethal. During the 20th century, biologists redefined the species concept in evolutionary terms and we now look on a species as a group of organisms that can interbreed with one another. If anything, this is more problematic with microorganisms because there are a variety of methods by which genes can be exchanged between prokaryotes that according to their biochemical and physiological properties are different species. The barrier to gene flow that is central to the species concept therefore does not hold with prokaryotes.

Genome sequencing has emphasized the difficulties in applying the species concept to prokaryotes. It has become clear that different strains of a single species can have very different genome sequences, and may even have individual sets of strain-specific genes. This was first shown by a comparison between two strains of *Helicobacter pylori*, which causes gastric ulcers and other diseases of the human digestive tract. The two strains were isolated in the UK and USA and have genomes of 1.67 Mb and 1.64 Mb, respectively. The larger genome contains 1552 genes and the smaller one 1495 genes, 1406 of these genes being present in both strains. In other words, some 6-7% of the gene content of each strain is unique to that strain. A much more extreme distinction between strains was revealed when the sequence of the common laboratory strain of *E. coli*, K12, was compared with that of one of the most pathogenic strains, O157:H7. The lengths of the two genomes are significantly different – 4.64 Mb for K12 and 5.53 Mb for O157:H7 – with the extra DNA in the pathogenic strain scattered around the genome at almost 200 separate positions. These 'O-islands' contain 1387 genes not present in *E. coli* K12, many of these genes coding for toxins and other proteins that are clearly involved in the pathogenic properties of O157:H7. But it is not simply a case of O157:H7 containing extra genes that make it pathogenic. K12 also has 234 segments of its own unique DNA, and although these 'K-islands' are, on average, smaller than the O-islands, they still contain 528 genes that are absent from O157:H7. The situation therefore is that *E. coli* O157:H7 and *E. coli*K12 each has a set of strain-specific genes, which make up 26% and 12% of the gene catalogs, respectively. This is substantially more variation than can be tolerated by the species concept as applied to higher organisms, and is difficult to reconcile with any definition of species yet devised for microorganisms.

The difficulties become even more acute when other bacterial and archaeal genomes are examined. Because of the ease with which genes can flow between different prokaryotic species, it was anticipated that the same genes would occasionally be found in different species, but the extent of lateral gene transfer revealed by sequencing has taken everybody by surprise. Most genomes contain a few hundred kb of DNA acquired directly from a different species, and in some cases the figure is higher: 12.8% of the *E. coli* K12 genome, corresponding to 0.59 Mb, has been obtained in this way. A second surprise is that transfer has occurred between very different species, even between bacteria and archaea. For example, the thermophiles bacterium *Thermatoga maritima* has 1877 genes, 451 of which appear to have been obtained from arch eons. Transfer in the other direction, from bacteria to archaea, is equally prevalent. The picture that is emerging is one in which prokaryotes living in similar ecological niches exchange genes with one another in order to increase their individual fitness for survival in their particular environment. Many of the *Thermatoga* genes that have been obtained from arch eons have probably helped this bacterium acquire its ability to tolerate high temperatures.

Lateral gene transfer has clearly played an important role in the evolution of prokaryotes. Unlike higher organisms, the evolutionary histories of bacteria and archaea cannot be described as a simple branching pattern, but instead have to incorporate the horizontal flow of genes between species. As well as its impact on our understanding of evolution, lateral gene transfer also has implications for the way in which evolutionary relationships are inferred by molecular phylogenetics. With higher organisms, comparisons between the sequences of equivalent genes in different species can be used to reconstruct the evolutionary relationships between those species. This type of analysis assumes that evolution has followed the simple branching pattern illustrated in, and cannot, therefore, be used to infer relationships between prokaryotes if there is a chance that the genes being analyzed have been transferred laterally between any of the species being studied. But the analysis was used for many years before the extent of lateral gene transfer was recognized and microbiologists are now faced with the need to reassess the validity of the evolutionary schemes that were established in the pre-genome era.

Speculation on the Minimal Genome Content and the Identity of Distinctiveness Genes

Even though a number of prokaryotic genome sequences have now been published, it is not yet possible to describe a complete catalog of the gene content of any one species, for the simple reason that the functions of many of the genes are unknown. Over 1500 of the *E. coli* K12 genes, for example, have not yet been assigned a function. Despite the incompleteness of the information, it is still interesting to examine the roles of the genes whose functions are known, and to appreciate the number of different genes involved in each of the various biochemical activities that a bacterium such as *E. coli* is able to carry out.

Gene catalogs are even more interesting when comparisons are made between different species. We see, for example, that whereas 243 of the identified genes in the *E. coli* genome are involved in energy metabolism, *Haemophilus influenzae* has only 112 genes in this category and *Mycoplasma genitalium* just 31. These comparisons have led to speculation about the smallest number of genes needed to specify a free-living cell. Theoretical considerations initially led to the suggestion that 256 genes are the minimum required, but experiments in which increasing numbers of *Mycoplasma* genes have been mutated suggest that 265-350 are needed. There has been similar interest in searching for 'distinctiveness' genes – ones that distinguish one species from another. Of the 470 genes in the *M. genitalium* genome, 350 are also present in the distantly related bacterium *Bacillus subtilis*, which suggests that the biochemical and structural features that distinguish a *Mycoplasma* from a *Bacillus* are encoded in the 120 or so genes that are unique to the former. Unfortunately, the identities of these supposed distinctiveness genes do not provide any obvious clues about what makes a bacterium a *Mycoplasma* rather than anything else.

Eukaryotic Genomes

The genome sizes of eukaryotes are tremendously variable, even within a taxonomic group (so-called C-value paradox). Eukaryotic genomes are divided into multiple linear chromosomes; each chromosome contains a single linear duplex DNA molecule. Eukaryotic genes in a biochemical or signaling pathway are not organized into operons; one mRNA makes one protein.

A typical eukaryotic gene, therefore, consists of a set of sequences that appear in mature mRNA (called exons) interrupted by introns. The regions between genes are likewise not expressed, but may help with chromatin assembly, contain promoters, and so forth.

Intron sequences contain some common features. Most introns begin with the sequence GT (GU in RNA) and end with the sequence AG. Otherwise, very little similarity exists among them. Intron sequences may be large relative to coding sequences; in some genes, over 90 percent of the sequence between the 5′ and 3′ ends of the mRNA is introns. RNA polymerase transcribes intron sequences. This means that eukaryotic mRNA precursors must be processed to remove introns as well as to add the caps at the 5′ end and polyadenylic acid (poly A) sequences at the 3′ end.

Eukaryotic genes may be clustered (for example, genes for a metabolic pathway may occur on the same region of a chromosome) but are independently controlled. Operons or polycistronic mRNAs do not exist in eukaryotes. This contrasts with prokaryotic genes, where a single control gene often acts on a whole cluster (for example, *lac*I controls the synthesis of β-galactosidase, permease and acetylase).

One well-studied example of a clustered gene system is the mammalian globin genes. Globins are the protein components of hemoglobin. In mammals, specialized globins exist that are expressed in embryonic or fetal circulation. These have a higher oxygen affinity than adult hemoglobins and thus serve to "capture" oxygen at the placenta, moving it from the maternal circulation to that of the developing embryo or fetus. After birth, the familiar mature hemoglobin (which consists of two alpha and two beta subunits) replaces these globins. Two globin clusters exist in humans: the alpha cluster on chromosome 16, and the beta cluster on chromosome 11, these clusters, and the gene for the related protein myoglobin, probably arose by duplication of a primordial gene that encoded a single heme-containing, oxygen-binding protein. Within each cluster is a gene designated with the Greek letter Ψ. These are pseudo genes—DNA sequences related to a functional gene but containing one or more mutations so that it isn't expressed.

The information problem of eukaryotic gene expression therefore consists of several components: gene recognition, gene transcription, and mRNA processing. These problems have been approached biochemically by analyzing the enzyme systems involved in each step.

Genome Size and Complexity

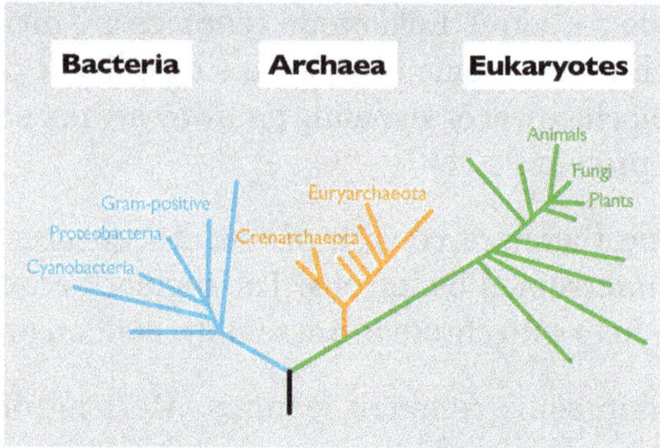

Phylogenetic relationship between members of the three domains of cellular life.

The genome and the complexity of living beings, the genome of an organism is the whole DNA content of its cells, including genes and intergenic regions. In prokaryotes (Archaea and Bacteria) there is, in general, a linear relationship between genome size and the number of genes. The smallest genomes are found in symbionts and parasites, as they undergo a gene degradation process during adaptation to their new lifestyle. However, in eukaryotes there is no correlation between genome size and the complexity of the organism. This is known as the C-value paradox. The largest genome is found in an amoeba, a one-cell organism, with 686,000 Mb, 200 fold larger than the human genome and 20,000 fold larger than the one found in yeast.

Figure below represents the range of sizes of genome found in the three domains of life: bacteria, archaea and eukaryotes. It seems clear that prokaryotes are, in general, smaller than eukaryotes, with the exception of some large-sized bacteria and some very small-sized eukaryotes.

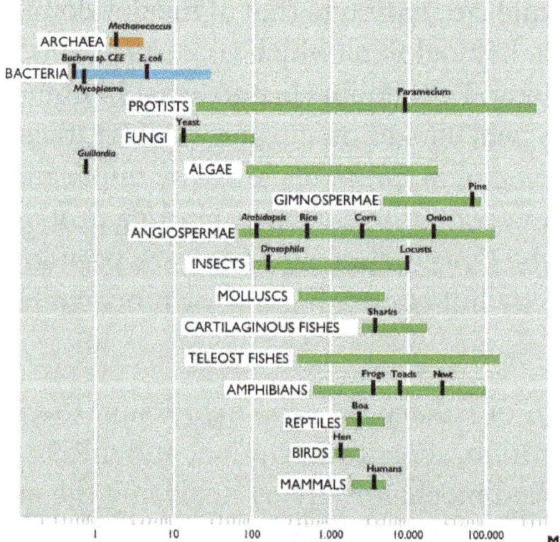

Range of genome size in organisms of the three domains of life.

Prokaryotes: Bacteria and Archaea

According to the data published so far, the size varies from 0.58 mega bases (1 mega base (Mb) is one million base pairs (bp)) in the intracellular pathogen Mycoplasma genitalium, to more than 10 Mb in several species of cyanobacteria, with the exception of Bacillus megaterium, which has a genome of 30 Mb. The second smallest genome ever published is that of Buchnera sp. APS, endosymbiont of the cereal aphid Acyrthosiphon pisum, with a size of 641 kb. Recently, a research group has characterized six genomes smaller than even those of Mycoplasma, the smallest of all being that of Buchnera sp. CCE, endosymbiont of the aphid Cinara cedri, with a size of 0.45 Mb. In general, most genomes are less than 5 Mb in size.

Is there a relationship between genome size and number of genes? The size of the prokaryotic gene is uniform, about 900 to 1000 bp. Therefore, one can estimate the gene density at each sequenced genome.

We can conclude that, at least in prokaryotes, genomes have a larger number of genes and are also more complex. That is, the number of genes reflects the lifestyle. Thus, smaller bacteria are specialists, such as obligate parasites and endosymbionts, and larger bacteria are generalists, and may even have a certain degree of development, such as sporulation in Bacillus.

Eukaryotes: C-Value Paradox

Genome size in eukaryotes is defined as the C-value or amount of DNA per haploid genome, such as that which exists in the nucleus of a spermatozoon. It is called C, for constant or characteristic, to indicate the fact that size is practically constant within a species.

Referring back to figure above, we see that, in general, eukaryotes have larger genomes than prokaryotes, except for some endosymbiont or parasitic green algae, which have very small genomes. Specifically, the smallest eukaryotic genome ever sequenced is that of Guillardia theta, symbiont red algae, of only 0.55 Mb. We can also see in the figure that there is a wide range of sizes, much greater than that of prokaryotes, more than 80,000-fold larger, from organisms such as yeast (1.2 Mb) to the amoeba (686,000 Mb). But is there, as in bacteria, a relationship between genome size and complexity of the organism?

In figure above we have represented the range of C-value in several representative groups of eukaryotic organisms. As we can observe, unicellular protists such as amoebae show the greatest variation in C-values (23.5 Mb to 686.000 Mb, with a ratio of 29,191 between the largest and the smallest) while mammals, birds and reptiles show less variation in the size of their genome (a ratio of only 4, 1 and 4, respectively). Furthermore, the large variation in genome sizes between eukaryotic species does not seem to have a relationship with either the complexity of the organism or the number of genes they contain. For example, amoebae, which have the largest genomes, have 200 times more DNA than humans (3,400 Mb) and it is clear that an amoeba cannot be more complex than a human. Moreover, it would be expected that mammals, more complex organisms, present larger genomes. However, many other organisms, such as fish, amphibians or plants, have much larger genomes. Even when we compare the sizes between organisms that appear similar in terms of complexity, there are also wide differences in their C-values. To give some examples, flies and locusts, onions and lilies, etc. have considerable variations in the sizes of their genomes. Amphibians as a group have variations of up to 91 times and it is hard to believe that this may reflect variations of nearly 100 times the number of genes necessary to give rise to the corresponding amphibians, or that onions need 200 times more DNA than rice. Figure below shows some living beings with size proportional to the size of their genome and needs no further explanation.

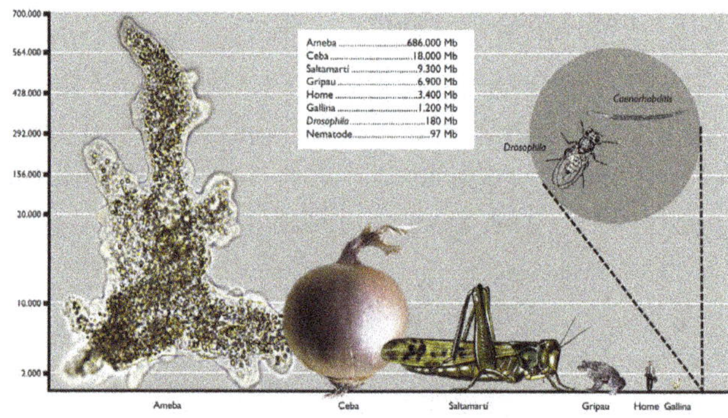

Genome size in some living beings. The height of the drawings is proportional to the size of their genome. The specimes are amoebae, onions, grasshoppers, toads, humans, hens, Drosophila and Caenorhabditis, a nematode worm.

The mismatch between the C-values and the presumed amount of genetic information contained within the genomes was called C-value paradox. Since we cannot assume that a species possesses less DNA than the quantity required specifying its vital functions, we have to explain why many species contain this amount of excess DNA.

Gene Dna Or Non-Gene Dna

The first question that has to be clarified is whether there is a correlation between genome size and the number of genes. That is, are the differences in genome sizes due to gene or non-gene DNA?

We have known since the late 60s that the eukaryotic genome is composed of a large amount of repetitive DNA. Moreover, since the late 70s we have known that genes are interrupted by non-coding sequences, introns, which must be removed before the ribosome synthesizes protein. We are talking in both cases about a seemingly superfluous DNA, which contributes to the wide variation in C-values and therefore explains the apparent paradox.

The size and number of introns vary widely along the evolutionary scale, mammals being the ones with the highest number and larger size. Repetitive DNA also varies between organisms. Traditionally this DNA is classified as: highly repetitive, with sequences such as microsatellites and mini satellites; and moderately repetitive, where transposable elements, the sequences that constitute the clearest example of selfish DNA, are found.

Number of Genes and Complexity of the Organism

As sequences of whole genomes are completed, we will know with more or less accuracy the number of genes derived from these sequences, since what we had so far were indirect estimates. However, some data is proving to be surprising because, in some cases, there appears to be a clear correlation between the number of genes and the complexity of the organism. The nematode worm C. elegans has 18,000 genes, about 5,000 more than Drosophila, a more complex organism. Man has only twice as many genes as C. elegans (estimates indicated about 100,000). We are also beginning to understand these data. There are mechanisms in higher eukaryotes that are able to "expand the proteome". That is, from the same DNA sequence, they can obtain more than one protein. Large introns found in mammals can, in many cases, "hide" information that cannot be inferred only with the DNA sequence. It will be some time before we can determine the number of proteins that an organism is able to synthesize. But this would be the subject of another paper. In any case, we can say that more complex organisms have more gene functions.

Table: Genome size, gene number and gene density.

Organism				
Common name or class	Scientific name	Genome size (Mb)	Number of genes	Gene density (genes/Mb)
Eukaryotes	*Saccharomyces cerevisiae*	12	6,241	480
Nematode	*Caenorhabditis elegans*	97	18,424	190
Cruciferous	*Arabidopsis thaliana*	125	25,498	204
Fruit fly	*Drosophila melanogaster*	180	13,601	75
Pufferfish	*Fugu rubripes*	400	35,000	100
Rice	*Oryza sativa*	450		
Sea urchin	*Strongylocentrotus purpuratus*	900	27,350	30
Maize	*Zea mays*	2,400		
Human	*Homo sapiens*	3,400	35,000	10
Onion	*Allium cep*	18,000		
Amoeba	*Amoeba dubia*	686,000		
Archaea	*Aeropyrum pernix*	1.55	1,522	981

Euryarchaeota	*Methanococcus jannaschii*	1.66	1,715	1033
Euryarchaeota	*Archaeoglobus*	2.18	2,420	1110
Bacteria	*Buchnera sp. CCE*	0.45		
Gram positive	*Mycoplama genitalium*	0.58	479	831
Proteobacteria	*Buchnera sp. APS*	0.64	564	881
Gram negative	*Haemophilus influenzae*	1.8	1,727	959
Cyanobacteria	*Synechocystis sp.*	3.6	3,168	880
Gram positive	*Bacillus subtilis*	4.2	4,100	976
Proteobacteria	*Escherichia coli*	4.6	4,288	932

Molecular Mechanisms that Alter Genome Size

There are many mutational mechanisms that can produce changes in genome size. Some of them occur on a large scale (whole genome duplication), while others occur on a very small scale (loss or gain of a few nucleotides). However, we must note that these mutations affect, in principle, a single cell, and if it is a gamete, the mutation may be transmitted to an offspring. This individual will have to live with others in a population and only in future generations will we know if the mutation it carries will extend to all individuals in the population (fixation) or, on the contrary, will disappear. The probability of one or the other occurring depends on evolutionary mechanisms such as natural selection (whether it provides an advantage or disadvantage to the individual) or genetic drift (random).

Chromosomal mechanisms often produce drastic changes with a single mutation. We can highlight the whole genome duplication, duplication affecting a single chromosome, or part of it. Equivalently, mutations that result in the loss of some chromosome fragment are also known. Such changes, though also frequent in plants and animals, appear to have significantly contributed to shape the evolution of the genome of the former.

Complex Organisms have more Gene Functions

The mobile genetic elements or transposable elements are other causes of large variations in genome size. These elements, of a few thousand nucleotides, are duplicated and the duplicate copies are inserted in other parts of the genome, causing rapid increases in its size, unless the mechanisms that control their proliferation intervene (as authentic selfish elements, they develop mechanisms to regulate their movement in the genome, so as not to produce irreparable damage, since that would imply their own disappearance). It has been estimated that the maize genome has been duplicated because of the transposable elements in the last three million years of its evolution.

It is considered that the spontaneous insertions or deletions (called indels) of a few nucleotides are one of the most important causes of the development of the size of the genome on the long term. In several species of insects, for example, a strong correlation has been shown between the overall rate of DNA loss of intergenic and non-coding regions and genome size. The fixation of these mutations is very unlikely if the indel affects a gene, but it is more likely if it affects pseudo genes (nonfunctional genes, recently inactivated) or other nonfunctional DNA sequences. The disintegration of the genes, or the disappearance of a gene in the genome, usually occur in a first step with its inactivation by a point mutation (formation of a pseudo gene). Subsequently, the DNA that forms the pseudo genes is progressively removed until it disappears completely from the genome.

The variation in the length of the DNA of the mini satellites and microsatellites is another mechanism which can alter the size of the genome. These sequences are formed by a unit of few nucleotides, repeated contiguously from less than ten to several thousand times. The number of repeated units varies greatly, even in individuals of the same species. This is due to two mechanisms that cause both an increase and a decrease in the number of repetitions. These mechanisms are the recombination, called unequal, and the errors during DNA replication, due to the phenomenon known as slippage of the DNA polymerase. Indeed, these are the sequences that are being used for the identification of human remains, for example, due to their high variability.

One of the most important processes in the increase of genome size in unicellular organisms, especially prokaryotes, is horizontal gene transfer. This process involves the introduction into the genome of a DNA fragment from another species, containing one or more genes. The DNA is introduced into the cell by different mechanisms and must then recombine with the genome. If the genes introduced confer some advantage to the organism they remain like that, otherwise they can mutate and become inactive. An example of the importance of these processes can be seen in the intestinal bacteria Escherichia coli, some strains of which have become pathogenic for the transfer of virulence genes. The significance of this phenomenon in recent periods of evolution of higher organisms is much more limited, but there are many reported cases of intracellular DNA gene transfer from the genome of the mitochondria and chloroplasts to the nuclear genome.

All these mutations arise periodically in many species, including humans. Many of them are disadvantageous and selective pressure progressively removes them from the population, others are individually neutral but may be collectively beneficial or harmful, by increasing or decreasing the size of the genome. In bacteria, having a small genome can be positive to optimize the time and cost of DNA replication. In eukaryotic organisms, several advantages have been proposed for having a large genome size, though this is not always associated with a higher number of genes.

Coding Regions

The coding region of a gene is the part of the gene that will be eventually transcribed and translated into protein, i.e., the sum total of its exons. The remaining portion of the gene is interspersed by introns (see the figure below), or regions that are trimmed away during RNA splicing and thrown out. Meanwhile, the gene's exons are assembled into a single strand of mRNA for translation into protein.

Normally, only a small percentage of a given gene will actually form its coding regions. In fact, only about 5% of the human genome belongs to coding regions; in some eukaryotes, this percentage is as low as 1%.

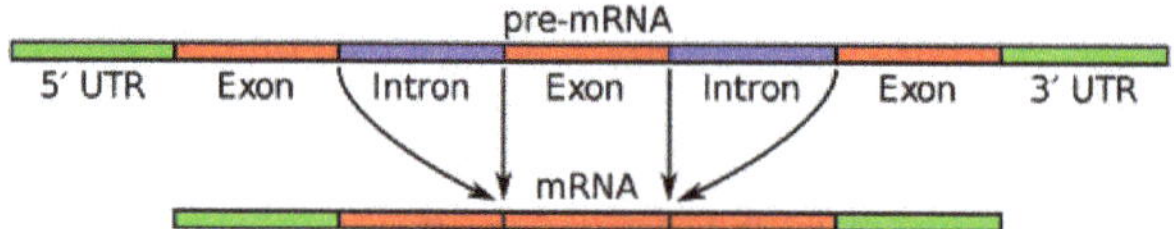

Coding regions often correspond to GC-rich areas of the genome or areas where the ratio of transitions to transversions is elevated.

Different Coding Regions

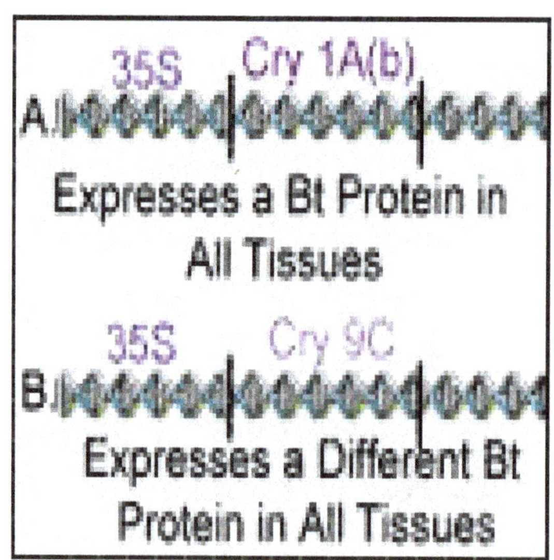

Changing the coding region of a gene will result in the production of a different protein

Many Bt genes have been cloned that have the potential to provide resistance to ECB and other insect pests. Three coding regions for ECB resistance have been used in commercial Bt corn; Cry 1A(b), Cry 1A(c), and the Cry 9c. Each region encodes crystalline proteins in the bacteria that are responsible for insect larvae toxicity. When eaten by the European corn borer, these crystalline, or Cry proteins, bind to the insects' midgut causing those cells to burst from a water imbalance killing the corn borer. Transgenic plants containing a Bt gene produce these Cry proteins, which the ECB ingest when feeding on the plants. There are hundreds of naturally occurring Bt proteins each with their own coding regions. However, not all are toxic to ECB.

The nucleotide sequences of the Cry 1A(b) and Cry 1A(c) coding regions are very similar. The slight differences are not enough to cause a difference in protein toxicity. The nucleotide sequence of the Cry 9C coding region, on the other hand, is different enough that it produces a protein that is toxic to European corn borer, but binds to a different site in the midgut killing the corn borer larvae in a slightly different way. Therefore even if ECB become resistant to the Cry 1A(b) and Cry 1A(c) proteins they may still be susceptible to the Cry 9(c) protein.

This technique has potential for dealing with the development of insect resistance to the toxin. However, it is not a "silver bullet". It is possible that when ECB develops resistance to one Bt protein, it will also develop resistance to other Bt proteins. This is called cross-resistance. The alterations of coding regions to avoid resistance requires an understanding of how the insecticidal proteins interact with the insect on a biochemical level.

Exome

The exome is a subset of the genome that contains many of the most important DNA sequences: the portions of genes that encode proteins. Most genes do not encode proteins using a single continuous stretch of DNA. Typically, a gene will encode a protein using multiple distinct stretches of DNA; the information in these separated stretches of DNA is then pieced together to direct production of a protein.

The stretches of the gene that are used to direct protein production are called "exons". Exons are also referred to as the coding region of a gene because they encode the information for a protein. The stretches of genes that do not encode proteins are called "introns" and are one kind of noncoding DNA sequence. The exons of all our genes make up approximately 1.5% of our genome and are collectively referred to as the "exome".

Because the exome is a small percentage of the entire genome, it is cheaper and faster to sequence the exome then the entire genome. While the exome contains many of the most important DNA sequences for understanding biological processes, there are some important DNA sequences that are not contained within the exome. There are regions of noncoding DNA that have important biological functions, such as regulating the coding regions of the genome. Ongoing research is investigating the functions of different regions of noncoding DNA.

References

- Virus-genomes, microorganisms-viruses: shmoop.com, Retrieved 27 June 2018
- Viral-genomes-chromosome-35972: biologydiscussion.com, Retrieved 28 April 2018

- Eukaryotic-gene-structure, eukaryotic-genes, biology: cliffsnotes.com, Retrieved 14 July 2018
- The-size-of-the-genome-and-the-complexity-of-living-beings, monographs: metode.org, Retrieved 14 March 2018
- Gene-coding-region, glossary: rosalind.info, Retrieved 15 May 2018
- Genome-43: nature.com, Retrieved 17 April 2018

Chapter 3

Understanding DNA and RNA

DNA or deoxyribonucleic acid is a double helix structure that carries the genetic instruction useful for the growth, functioning and reproduction of organisms. RNA or ribonucleic acid is a molecule essential for the regulation, coding, decoding and expression of gene. The aim of this chapter is to provide an easy understanding of DNA and RNA. It elucidates topics like DNA condensation, DNA fragmentation, DNA methylation, RNA interference, RNA processing, etc.

DNA

Deoxyribonucleic acid or DNA is a molecule that contains the instructions an organism needs to develop, live and reproduce. These instructions are found inside every cell and are passed down from parents to their children.

DNA Structure

DNA is made up of molecules called nucleotides. Each nucleotide contains a phosphate group, a sugar group and a nitrogen base. The four types of nitrogen bases are adenine (A), thymine (T), guanine (G) and cytosine (C). The order of these bases is what determines DNA's instructions, or genetic code. Human DNA has around 3 billion bases, and more than 99 percent of those bases are the same in all people, similar to the way the order of letters in the alphabet can be used to form a word, the order of nitrogen bases in a DNA sequence forms genes, which in the language of the cell, tells cells how to make proteins. Another type of nucleic acid, ribonucleic acid, or RNA, translates genetic information from DNA into proteins.

Nucleotides are attached together to form two long strands that spiral to create a structure called a double helix. If you think of the double helix structure as a ladder, the phosphate and sugar molecules would be the sides, while the bases would be the rungs. The bases on one strand pair with the bases on another strand: adenine pairs with thymine, and guanine pairs with cytosine.

DNA molecules are long — so long, in fact, that they can't fit into cells without the right packaging. To fit inside cells, DNA is coiled tightly to form structures we call chromosomes. Each chromosome contains a single DNA molecule. Humans have 23 pairs of chromosomes, which are found inside the cell's nucleus.

DNA Discovery

DNA was first observed by a German biochemist named Frederich Miescher in 1869. But for many years, researchers did not realize the importance of this molecule. It was not until 1953 that James Watson, Francis Crick, Maurice Wilkins and Rosalind Franklin figured out the structure of DNA — a double helix — which they realized could carry biological information.

Watson, Crick and Wilkins were awarded the Nobel Prize in Medicine in 1962 for their discoveries concerning the molecular structure of nucleic acids and its significance for information transfer in living material. Franklin was not included in the award, although her work was integral to the research.

DNA Sequencing

DNA sequencing is technology that allows researchers to determine the order of bases in a DNA sequence. The technology can be used to determine the order of bases in genes, chromosomes, or an entire genome. In 2000, researchers completed the first full sequence of the human genome.

DNA Testing

A person's DNA contains information about their heritage, and can sometimes reveal whether they are at risk for certain diseases. DNA tests or genetic tests, are used for a variety of reasons, including to diagnose genetic disorders, to determine whether a person is a carrier of a genetic mutation that they could pass on to their children, and to examine whether a person is at risk for a genetic disease. For instance, mutations in the BRCA1 and BRCA2 genes are known to increase the risk of breast and ovarian cancer, and analysis of these genes in a genetic test can reveal whether a person has these mutations.

Genetic test results can have implications for a person's health, and the tests are often provided along with genetic counseling to help individuals understand the results and consequences of the test.

A-DNA

A-form DNA was first identified from fibre-diffraction studies of DNA at 'low' (75%) relative humidity. More recently, crystal studies have identified specific sequences which can adopt A-DNA type of structures. In general, A-DNA for any sequence is favored under dehydrating conditions, and certain purine stretches will favor an A-conformation, even in cases of higher hydration levels. It appears that at least four purines (or pyrimidine's) in a row are enough to set up a local A-DNA helix, although of course certain purine stretches are more likely to form A-DNA than others. (For example, the sequence AAAA crystallizes as B-DNA, not in the A helix.) It is thus possible to have a

DNA sequence that contains some regions in the A-form within the context of a mainly B-conformation.

Some of the helical parameters of A-DNA are given in table below. The A-DNA helix is a bit wider than B-DNA (and also Z-DNA), and this is mainly due to the fact that the base pairs stack nearly on top of each other in B-DNA, but stack a little off-Centre in the A-conformation. A-DNA helix is less stable than the B-DNA conformation. A-DNA is also more rigid than B-DNA, again because the off-Centre stacking of the bases makes them less flexible. There are about 11 bp per turn for ADNA, compared with about 10 bp per turn for the B-form. Finally, the base-pair tilt is higher in A-DNA than in BDNA. A-helix is the common form for DNA–RNA hybrids, as well as double-stranded RNA; this is due to the extra OH group on the ribose sugar, which cannot fit easily into the tight space allotted to it in B-DNA.

Table: Comparison of different helical parameters for A-, B-, and Z-DNA

Parameter	A-DNA	B-DNA	Z-DNA
Helix sense	Right	Right	Left
Base pairs per turn	11	10	12
Axial rise (nm)	0.26	0.34	0.45
Helix pitch (°)	28	34	45
Base pair tilt (°)	20	-6	7
Twist angle (°)	33	36	-30
Diameter of helix (nm)	2.3	2.0	1.8

Biology of A-DNA

A-form helices are common for DNA–RNA hybrids, as well as for double-stranded RNA; in addition, the A-conformation is favored in triplex DNA. A transition from B-DNA to A-DNA has been postulated to occur during transcription, where the RNA–DNA hybrid would be more stable in the A-conformation. A-DNA also plays a role in some processes that do not involve RNA. For example, in sporulation bacteria, there is a protein which can bind to DNA in the B-conformation and induce a change to the A-DNA helix. Another common biological occurrence of sequences which can readily form A-DNA is in the long terminal repeats (LTRs) of transposable elements. These regions often contain purine stretches which favor the A-DNA conformation. In fact, the DNA sequence used for the crystal structure sequence of A-DNA shown in figure is from an LTR of the human immunodeficiency virus. It is likely that these regions are involved in recombination. Short stretches of purines which are likely to form A-DNA conformations exist in genomes in much greater abundance than would be expected from the mononucleotide composition, ranging from about a fourth of the genome in bacteria to close to half the DNA in eukaryotic chromosomes.

B-DNA

The B-DNA is the most common and predominate type of structural conformation of DNA in the cells. The DNA prefers to occur in B form under the natural physiological conditions (pH and salt concentration) in the cell. The B-DNA is better described as the Watson – Crick Model of DNA described for the first time by James Watson and Francis Crick. Important structural features of B-DNA are given below:

- Majority of the DNA in a cell is in B-DNA conformation.
- B-DNA is a right handed helix.
- In B-DNA, the bases occupy at the core whereas the sugar phosphate backbone occurs at the peripheral portion of the helix.
- In B-DNA only the edges of the base pairs are exposed to the solvent.
- Each base pair in B-DNA has the same width.
- The width of A – T and G – C in B-DNA is 10.85 Å.
- The helical diameter of B-DNA is 20 Å.
- Each turn on helix in B-DNA possess a helical height of 34 Å.
- Each turn in the B-DNA consists of 10 base pairs.
- The distance between adjacent base pairs in B-DNA is 3.4 Å.
- Each base pair will have a helical twist of 36° (360/10).
- The plain of inter-strand hydrogen bonds are perpendicular to the helical axis.
- B-DNA has a solid central core.
- The major grove of B-DNA is wide and deep.
- The minor grove of B-DNA is narrow and deep.
- The sugar pucker in B-DNA is C2' endo form.
- The glycosidic bond conformation in B-DNA is in anti- form.

Z-DNA

Z-DNA is a left-handed double helical conformation of DNA in which the double helix winds to the left in a zigzag pattern. The DNA strand with complementary nucleotides with alternating purines and pyrimidines (such as poly-d(GC).poly-d(GC) or polyd(AC).poly-d(GT)) can form Z DNA conformation at high salt concentration. The existence of Z DNA was discovered by Andres Wang and Alexander Rich. Z-DNA is one of the biologically active forms of DNA found in vivo in the cells. The exact biological function of Z-DNA is not clear. The Z-DNA is usually located upstream of the start site of a gene and thus it may have some role in the regulation of gene expression.

Formation of Z-DNA in Vitro

The existence of Z-DNA was first suggested by optical studies demonstrating that a polymer of alternating deoxyguanosine and deoxycytidine residues ($d(CG)_n$) produced a nearly inverted circular dichroism spectrum in a high salt solution. The physical reason for this finding remained a mystery until an atomic resolution crystallographic study of $d(CG)_3$ revealed a left-handed double helix, which maintained Watson-Crick base pairing. The Z-DNA helix is built from a dinucleotide repeat with the deoxycytidines in the *anti*-conformation while deoxyguanosines are in the unusual *syn*form. In Z-DNA, there is a single narrow groove that corresponds to the minor groove of B-DNA. There is no major groove. Instead, the "information"-rich residues that allow sequence-specific recognition of B-DNA lie exposed on the convex outer surface of Z-DNA. The transition from B-DNA to Z-DNA involves "flipping" the base pairs upside down. During this process, deoxycytidine remains in the *anti*-conformation because both the sugar and base rotate, while only the base of deoxyguanosine inverts, moving it into the conformation. As a consequence, the backbone follows a zigzag path, giving rise to the name Z-DNA. Z-DNA can form from B-DNA under physiological salt conditions when deoxycytidine is 5-methylated. The demonstration that Z-DNA formed under conditions of negative super helical stress raised considerable excitement as this brought the left-handed conformation within the realm of biology.

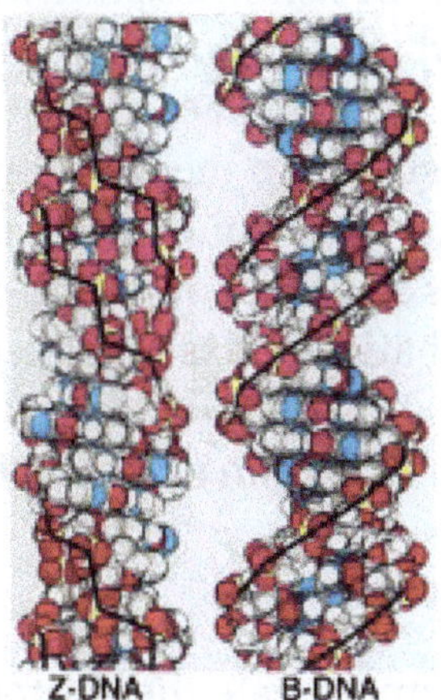

The "information-rich" residues that allow sequence-specific recognition of the major groove of B-DNA lie on the convex surface of left-handed Z-DNA helix. The two DNA strands of each duplex are highlighted by *solid black lines*. The "zigzag" nature of the Z-DNA backbone is clearly seen.

Stabilization of Z-DNA by negative supercoiling illustrates a number of features. First, Z-DNA is a higher energy conformation than B-DNA and will only form when plasmids are torsionally stressed. The energy necessary to stabilize Z-DNA can be determined by measuring the plasmid super helical density at which Z-DNA formation occurs, and it is proportional to the square of the number of negative supercoils. Second, sequences other than alternating purines and pyrimidine's can form Z-DNA. The ease with which this occurs depends on the sequence; d(CG)$_n$ is best, d(TG)$_n$ is next, and a d(GGGC)$_n$ repeat is better than d(TA)$_n$. Third, formation of B-Z DNA junctions, each of which has a free energy ΔG near +4 kcal/mol is a significant energetic barrier to Z-DNA formation.

Based on these empirical findings, computer models have been developed to rank the Z-DNA-forming potential of naturally occurring sequences. One analysis of 137 fully sequenced human genes demonstrated that sequences which could form Z-DNA easily were present in 98 and that they were distributed non randomly throughout the gene; sequences were 10 times more frequent in 5′ than in 3′ regions. This fits with the expectation that the energy necessary to form Z-DNA *in vivo* is generated by transcription. As demonstrated by Liu and Wang, negative supercoils arise behind a moving RNA polymerase as it ploughs through the DNA double helix. The torsional strain generated by passage of RNA polymerases then becomes a potent source of energy to stabilize Z-DNA.

Z-DNA in Prokaryotic Systems

A number of experiments have been used to demonstrate that Z-DNA can form *in vivo*. One approach uses chemical modification. Through use of either osmium tetroxide or potassium permanganate, it can be demonstrated that plasmids containing a d(CG)$_n$ insert will form Z-DNA *in vivo*. UV cross-linking of bacteria treated with psoralens have confirmed these results and made possible a precise measurement of the amount of unrestrained supercoiling present within *Escherichia coli*. A more sophisticated approach has used a construct in which an *Eco* RI site is embedded in a Z-DNA-forming sequence. In the bacterial cell, this fragment can be methylated when it is in the B-DNA conformation, but it becomes resistant to methylation while in the Z-DNA conformation. Susceptibility to methylation thus can be used as a measure of *in vivo* torsional strain. Results obtained with this system show that Z-DNA formation in *E. coli* occurs in the absence of external perturbation and is regulated by transcription, an effect that is enhanced by mutations inactivating topoisomerase I. Formation of Z-DNA, however, was not observed in *Morganella*, *Klebsiella* or *Enterobacter*.

Z-DNA in Eukaryotic Systems

It has been difficult to directly demonstrate the existence of Z-DNA in eukaryotic systems. A number of early observations clearly suggested its existence. Unlike B-DNA, Z-DNA is highly immunogenic, and polyclonal as well as monoclonal antibodies can be made that recognize this conformation. The first suggestion that Z-DNA was found in eukaryotic systems came from work with humans. Analysis of sera obtained from patients with autoimmune diseases, especially lupus erythematosis showed that these patients produced antibodies which were highly specific for Z-DNA. These were produced during the exacerbations of the disease, together with antibodies to many other nuclear components.

Antibodies raised in rabbits and sheep were used in staining experiments with both fixed and unfixed polytene chromosomes of *Drosophila*. These produced an unusual pattern with staining in the inter band regions but not in the bands. Staining was especially intense in the puffs, which are associated with high levels of transcriptional activity. Antibodies were also used in staining ciliated protozoa that have both a macronucleus and a micro nucleus. The micronucleus is used for genetic reproduction, but the macronucleus is the site of all transcriptional activity. Here, again, the macronucleus stained exclusively, with no staining in the micronucleus. Both of these early experiments suggested somewhat indirectly a link between transcriptional activity and the presence of Z-DNA.

Analysis of intact mammalian systems has been more complicated. There are a number of limitations in these experiments. No phenotype has been associated with the presence or absence of Z-DNA-forming sequences, thus limiting the use of genetic approaches. In addition, regulation of Z-DNA is likely to be very complex. For example, what is the importance of the three RNA polymerases relative to production of Z-DNA? It is known that RNA polymerase I works on some favorable Z-DNA-forming sequences in ribosomal RNA genes. In addition, it is not known how the torsional strain in regions 5′ to RNA polymerase II promoters is regulated. What is the influence of the TATA box sequence bound to its proteins? Are genes lacking a TATA box more topologically sensitive to the torsional strain generated by the moving polymerase? In this context, the effect of potential Z-DNA-forming sequences upstream of a promoter must be interpreted carefully; deletion or mutation of such regions, as in the case of the SV40 enhancer which has regions of alternating purine/pyrimidine repeats, may have many different interpretations.

A number of experiments have been carried out using metabolically active permeabilized mammalian nuclei, which were formed by embedding intact cells in agarose using the method of Jackson and Cook. Here, low concentrations of detergent are used to lyse the cytoplasmic membrane and permeabilize the nuclear membrane. These nuclei have been shown to replicate DNA at 85% of the rate observed in the intact cell, and they are active in transcription. In these experiments the amount of Z-DNA present in the gene is measured by diffusing biotin-labeled anti-Z-DNA monoclonal antibodies into the beads. The amount of Z-DNA was measured initially by the amount of radioactive streptavidin that would bind within the nucleus. These experiments showed that, at low concentrations of antibody, the amount of Z-DNA measured was independent of the antibody added over a 100-fold change in antibody concentration. Furthermore, the amount of Z-DNA depended on DNA negative torsional strain. It increased dramatically as transcription increased but was largely unaffected by DNA replication.

It was found that individual genes could be assayed by cross-linking the antibody to DNA using a 10-ns exposure of a laser at 266 nm. Release of DNA fragments with attached antibody was accomplished by diffusing in restriction endonucleases and performing an *in situ* DNA digest. Following isolation of biotin-labeled antibody-DNA complexes with streptavidin magneto beads, free DNA was obtained by proteolysis. These experiments made it possible to determine which regions of a gene form Z-DNA. Using hybridization or polymerase chain reaction techniques, the *c-myc*gene was studied in murine U937 cells. Three transcription-dependent Z-DNA-forming segments were identified in the 5′ region of the gene with two of them near promoters. Retinoic acid, which induces the cells to differentiate into macrophages, was then used to down-regulate expression of *c-myc*. Loss of *c-myc* expression was accompanied by a rapid reduction in the amount of Z-DNA present in these three regions. In contrast, Z-DNA was detectable by polymerase chain reaction with probes for the β-actin gene under all the conditions tested. β-Actin is not down-regulated with differentiation.

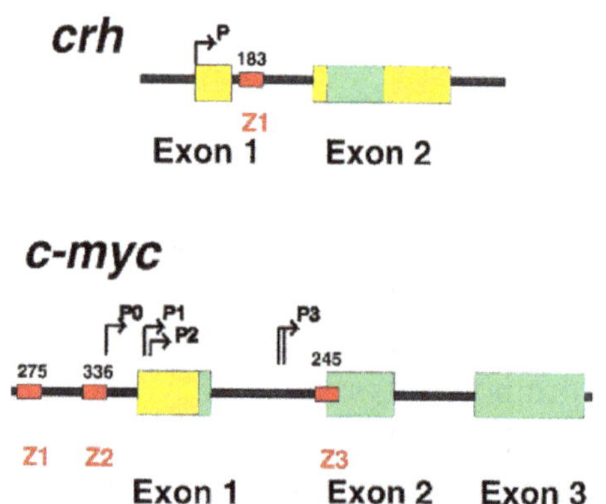

Z-DNA-forming segments, shown in *red*, can be detected in the genes encoding corticotropin releasing hormone (*crh*) and the *c-myc* 67-kDa protein product. Z-DNA forms only when these genes are transcriptionally active.

In other studies with a primary liver cell line, induction of Z-DNA was measured in the corticotropin hormone-releasing gene. Z-DNA formation increased when the gene was up-regulated and decreased when it was down-regulated. This finding suggests that physiological events are being measured in these systems.

A major conclusion from these studies is that Z-DNA forms largely, if not exclusively, behind a moving RNA polymerase and is stabilized by the negative supercoiling generated by DNA transcription.

Functional Consequences of Z-DNA Formation

In principle, Z-DNA formation could have a functional role that need not involve recognition of its shape by proteins. It has been shown that *E. coli* RNA polymerase does not transcribe through Z-DNA. Thus, the formation of Z-DNA behind (5′) a moving polymerase may block the following RNA polymerase from transcribing that region of a gene. This might ensure spatial separation between successive polymerases. As a consequence, processing of RNA would then be physically and temporally removed from that of subsequent transcripts, perhaps minimizing non-functional eukaryotic trans-splicing.

Alternatively, formation of Z-DNA could facilitate recombination of homologous chromosomal domains by relieving topological strain that arises when intact duplexes are intertwined. The Z-DNA-forming $d(CA/GT)_n$ sequence has been shown to be recombinogenic in yeast but is found to be less efficient than $d(CG)_n$ in human cells. Furthermore, there have been several reports correlating chromosomal breakpoints in human tumors to potential Z-DNA-forming sequences, although no causal relationship has yet been established. Last, Z-DNA formation could affect the placement of nucleosomes as well as the organization of chromosomal domains.

Identification of proteins that bind to Z-DNA would indirectly establish the presence of Z-DNA *in vivo* and help establish a biological role for this shape. There has been an extensive search by a number of laboratories for Z-DNA binding proteins. Early studies were unfruitful and caused widespread skepticism that Z-DNA would be associated with any biological function. Many of the positive results reported in these studies may have been due either to artifacts or misinterpretation of data. However, absence of proof was confused with absence of existence.

A High Affinity Z-DNA Binding Protein with Enzymatic Activity

Recent results give cause for optimism. An assay that by its design detects only proteins with high affinity for Z-DNA has revealed that one type of double-stranded RNA adenosine deaminase (dsRAD) called DRADA binds Z-DNA *in vitro*. The dissociation constant of the Z-DNA binding domain is nanomolar, making it likely that this interaction is of biological relevance. The domain maps to a region separate from the three copies of the RNA binding motif present in the protein and also from the catalytic domain.

DRADA is an example of a family of deaminases, the dsRAD family, that modify mRNA by catalyzing the hydrolytic deamination of adenine to inosine in regions that are double-stranded. RNA shape is important in this reaction as neither single-stranded RNA nor DNA are substrates for this reaction. The efficiency of editing *in vitro* is influenced by the length of dsRNA, with maximum efficiency seen in a synthetic substrate about 100 base pairs long. It is likely that different members of the dsRAD family will obtain specificity from the recognition of different RNA shapes. Members of the dsRAD family are ubiquitous in metazoa, suggesting that this activity is of great evolutionary significance. These enzymes may be an important source of phenotypic variation as they have the potential to significantly alter the linear flow of information from DNA to RNA. For example, inosine is translated as guanine, so that editing of a codon can result in the substitution of one amino acid for another. An illustrative example of the type of reaction that a dsRAD may catalyze is editing of the GluR-B receptor; whether DRADA or another member of the dsRAD family, such as RED-1, is involved in this reaction is currently a matter of debate. Editing of the second trans-membrane domain of the GluR-B receptor RNA results in the substitution of an arginine (CGG) for glutamine (CAG), changing the electrophysiological properties of the assembled receptor. The double-stranded RNA substrate that is modified by the enzyme is formed by folding the 3′-intron back onto the exon to base pair with the site that is edited. In this case, the involvement of introns requires that editing occurs soon after transcription of RNA and before splicing.

The potential involvement of introns in creating the substrates for editing provides a number of rationales for the recognition of Z-DNA by DRADA. Z-DNA *in vivo* is a transcription-dependent structure and will form when appropriate sequences are present behind (5′ to) a moving RNA polymerase. This transcription-induced Z-DNA may serve to localize DRADA to a particular region of a gene where editing is to occur, and it may also prevent indiscriminate editing of other regions. What is important is that the Z-DNA binding domain of the editing enzyme would target only transcribing genes and allow DRADA to act before the splicing apparatus removes the intron. In addition, recognition of Z-DNA by DRADA may block the gene from further transcription until editing of the RNA is complete. Currently there is no direct evidence *in vivo* that the Z-DNA binding domain influences the catalytic function of the enzyme. However, DRADA is present as a complex inside cells, and interactions may be mediated through other proteins. It should be possible to examine the role of Z-DNA recognition by DRADA *in vivo* by using UV cross-linking to identify the regions of genes that are bound to DRADA and correlate these with sites of dsRNA editing. It will also be of interest to determine whether other members of the dsRAD family are present within this complex. It is also possible that Z-DNA is not the only transcription-dependent structure recognized by this family of enzymes.

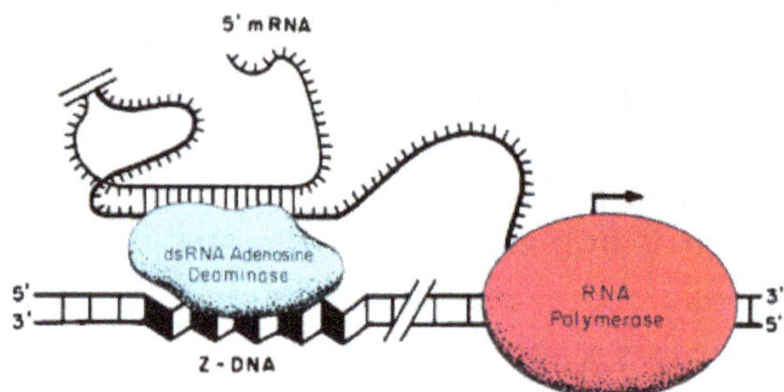

In vivo, Z-DNA is thought to be stabilized by the negative supercoiling generated by an RNA polymerase moving through a gene.

Transcription also gives rise to regions of double-stranded RNA (*dsRNA*), formed when a nascent RNA transcript folds back on itself. An RNA editing enzyme, dsRNA adenosine deaminase (DRADA), has been shown to bind both Z-DNA and dsRNA with nanomolar affinity. Each nucleic acid is bound by DRADA through a separate domain. This enzyme then catalyzes the hydrolytic deamination of adenine within the dsRNA to form inosine. Inosine is subsequently translated as guanine. Several editing sites may exist in a single pre-mRNA. DRADA thus utilizes the structural information encoded by DNA and RNA shapes to change the message read from a gene.

Other proteins may exist that bind to Z-DNA with lower affinity than DRADA. It has been demonstrated that peptides in which every second residue is lysine will stabilize

Z-DNA *in vitro* at micro-molar concentrations. This provides a simple protein motif with which to recognize Z-DNA. This motif exists in a number of proteins, but it remains to be shown that such proteins interact with Z-DNA. In addition, evidence has been presented to show that topoisomerase II from *Drosophila,* humans, and calf thymus recognizes a number of different DNA shapes, including Z-DNA.

Noncoding DNA

Only about 1 percent of DNA is made up of protein-coding genes; the other 99 percent is noncoding. Noncoding DNA does not provide instructions for making proteins. Scientists once thought noncoding DNA was "junk," with no known purpose. However, it is becoming clear that at least some of it is integral to the function of cells, particularly the control of gene activity. For example, noncoding DNA contains sequences that act as regulatory elements, determining when and where genes are turned on and off. Such elements provide sites for specialized proteins (called transcription factors) to attach (bind) and either activate or repress the process by which the information from genes is turned into proteins (transcription).

Noncoding DNA contains many types of regulatory elements:

- Promoters provide binding sites for the protein machinery that carries out transcription. Promoters are typically found just ahead of the gene on the DNA strand.
- Enhancers provide binding sites for proteins that help activate transcription. Enhancers can be found on the DNA strand before or after the gene they control, sometimes far away.
- Silencers provide binding sites for proteins that repress transcription. Like enhancers, silencers can be found before or after the gene they control and can be some distance away on the DNA strand.
- Insulators provide binding sites for proteins that control transcription in a number of ways. Some prevent enhancers from aiding in transcription (enhancer-blocker insulators). Others prevent structural changes in the DNA that repress gene activity (barrier insulators). Some insulators can function as both an enhancer blocker and a barrier.

Other regions of noncoding DNA provide instructions for the formation of certain kinds of RNA molecules. RNA is a chemical cousin of DNA. Examples of specialized RNA molecules produced from noncoding DNA include transfer RNAs (tRNAs) and ribosomal RNAs (rRNAs), which help assemble protein building blocks (amino acids) into a chain that forms a protein; microRNAs (miRNAs), which are short lengths of RNA that block the process of protein production; and long noncoding RNAs (lncRNAs), which are longer lengths of RNA that have diverse roles in regulating gene activity.

Some structural elements of chromosomes are also part of noncoding DNA. For example, repeated noncoding DNA sequences at the ends of chromosomes form telomeres. Telomeres protect the ends of chromosomes from being degraded during the copying of genetic material. Repetitive noncoding DNA sequences also form satellite DNA, which is a part of other structural elements. Satellite DNA is the basis of the centromere, which is the constriction point of the X-shaped chromosome pair. Satellite DNA also forms heterochromatin, which is densely packed DNA that is important for controlling gene activity and maintaining the structure of chromosomes.

Some noncoding DNA regions, called introns, are located within protein-coding genes but are removed before a protein is made. Regulatory elements, such as enhancers, can be located in introns. Other noncoding regions are found between genes and are known as inter genic regions.

The identity of regulatory elements and other functional regions in noncoding DNA is not completely understood. Researchers are working to understand the location and role of these genetic components.

Mitochondrial DNA

Mitochondrial DNA (mtDNA) is genetic material found in mitochondria. It is passed down from mothers to both sons and daughters, but sons cannot pass along their mothers' mtDNA to their children. This is because mtDNA is transmitted through the female egg. The mtDNA found in the egg is non-recombinant, meaning that it does not combine with any other DNA so that it is passed down virtually unchanged through the direct maternal line over the generations. You inherited your mtDNA exclusively from your mother.

The mtDNA test is more of an anthropological test than a genealogical test. From a genealogical standpoint, mtDNA is not very useful for two reasons. The first reason is mtDNA cannot be tied to any specific surname. Why? Because you inherited your mtDNA from your mother, who inherited it from her mother, who inherited it from her mother, and so on back in time. Traditionally, women in Western cultures change their surnames when they get married, so your mother did not have the same surname as her mother, who did not have the same surname as her mother, etc.

The second reason mtDNA is not very useful for historical genealogical research is because mtDNA mutates very, very slowly - much slower than yDNA. This means that your mtDNA is nearly identical to the mtDNA of your straight line maternal ancestor who lived thousands of years ago, and it is also identical to thousands of people living today. My mother has over 1,000 low resolution mtDNA matches in the FTDNA database. This means she shares a common maternal ancestor with them somewhere back in time. The problem is there is no way to know whether this common maternal ancestor lived recently or thousands of years ago.

In terms of recent kinship, mtDNA works best at disproving relationships rather than proving them. For example, if your mtDNA is not the same as your mother's mtDNA, this means she is not your biological mother and you were adopted. However, it is much more difficult (if not impossible) to tell from your mtDNA alone exactly who your biological mother really is, because your mtDNA is going to match the mtDNA of thousands of women.

Mitochondrial DNA Mutations

One cell contains numerous mitochondria, and each mitochondrion contains dozens of copies of the mitochondrial genome. Moreover, the mitochondrial genome has a higher mutation rate (about 100-fold higher) than the nuclear genome. This leads to a heterogeneous population of mitochondrial DNA within the same cell, and even within the same mitochondrion; as a result, mitochondria are considered heteroplasmic. When a cell divides, its mitochondria are partitioned between the two daughter cells. However, the process of mitochondrial segregation occurs in a random manner and is much less organized than the highly accurate process of nuclear chromosome segregation during mitosis. As a result, daughter cells receive similar, but not identical, copies of their mitochondrial DNA.

A nuclear gene, called DNA polymerase gamma (*POLG*), encodes the DNA polymerase responsible for replicating the mitochondrial genome.

The POLG protein consists of two domains: a catalytic domain that exhibits polymerase activity and an exonuclease domain that is involved in the recognition and removal of DNA base-pair mismatches that occur during DNA replication. A recent study suggests that mitochondria may have a nucleotide imbalance that leads to decreased POLG fidelity and higher mitochondrial DNA mutation rates.

Song and colleagues (2005) measured the mitochondrial levels of free deoxynucleotide triphosphates (dNTPs), the building blocks of new DNA strands made during DNA replication, in tissues from young and old rats. The researchers did not detect differences in the nucleotide levels in tissues taken from young versus old rats. However, they found that mitochondrial dNTP levels were highly divergent, and that dGTP was by far the most abundant nucleotide in the mitochondria of most tissues. In heart muscle and skeletal muscle, for example, they found that dGTP represented 85% to 91% of the mitochondrial nucleotide pool, whereas dTTP was present at 0.5%. Similarly, in the brain, dGTP represented 62% of the mitochondrial nucleotide pool, whereas dTTP constituted only 4%. In the liver, dGTP, dCTP, dATP and dTTP made up 37%, 51%, 9%, and 3% of the mitochondrial nucleotide pool, respectively. The abundant dGTP levels in mitochondria were in stark contrast to the dGTP levels in the rest of the cell: In whole rat embryos, dGTP represented only 10% of the free nucleotide pool.

DNA Testing to Determine Ancestry

A common public use for mtDNA in DNA testing is in determining ancestry. Because mtDNA does not change as rapidly as nuclear DNA, and because it is not mixed with the father's (paternal) DNA, it leaves a clearer record of distant ancestry – although only through the mothers' (maternal ancestry). Analysis of mtDNA is what allowed scientists to trace the maternal ancestry of all human beings to East Africa and to a time roughly between 100,000 and 200,000 years ago.

mtDNA testing can provide some kinds of information about ancestry not always apparent from regular DNA testing. For us non-scientists because mtDNA is shared among all people with the same maternal ancestry, it can provide some of the clearest proof of relationships among people with the same mother, grandmother, great grandmother, and so on.

Muscle Weakness

mtDNA testing is also important for diagnosing certain diseases. Mutations in the genes found in mtDNA have been shown to cause many different types of disease, often neuromuscular diseases or other diseases that cause muscle weakness. The three-time winner of the Tour de France, Greg Lemond gave up competitive cycling when he developed muscle weakness and fatigue. He was diagnosed with a mitochondrial disease called "mitochondrial myopathy" in 1994, and his New York Times story was perhaps the first to put mitochondrial diseases in the public eye.

Scientists have also determined that mutations in mtDNA are associated with symptoms and diseases of aging, leading to the theory that mitochondrial DNA changes are important for determining how rapidly people age. In the future, mtDNA testing could become an important part of determining a person's overall health. Even subtle changes in mtDNA may be important for determining your overall feeling of energy and well-being.

Mitochondrial Inheritance

In addition to chromosomal inheritance of nuclear DNA there is also mitochondrial inheritance, where specific characteristics are inherited exclusively from the mother. This does not mean that a descendant is automatically affected by a mitochondrial DNA disorder: that will depend on the quantity of mitochondrial DNA that carries the disorder.

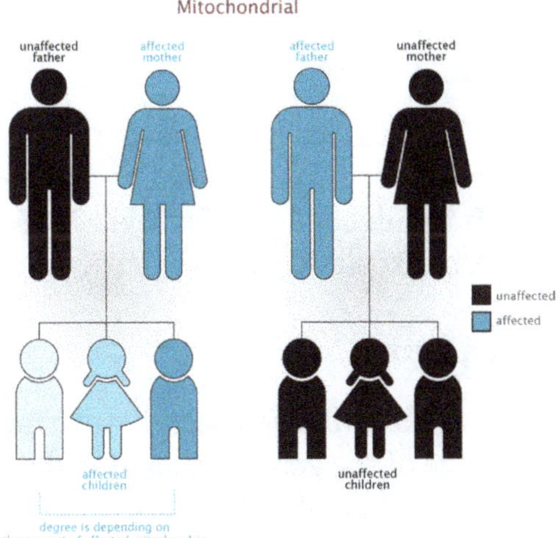

Every mitochondrial DNA-ring contains 37 genes. The genes are always the same, but the circular DNA molecules in a cell are not all identical. If they are, we call this homoplasmy; if not, it is called heteroplasmy. This is important in determining whether or not inherited mitochondrial disorders are expressed.

Every cell contains 100 to 1,000 mitochondria and each of these contains many mitochondrial DNA molecules. So if all the mitochondria in a cell contain the same (mutant or normal) DNA, this is a different situation from a mixture of normal and mutant mtDNA. The percentage of mutant mtDNA that is present determines the extent to which the condition is expressed.

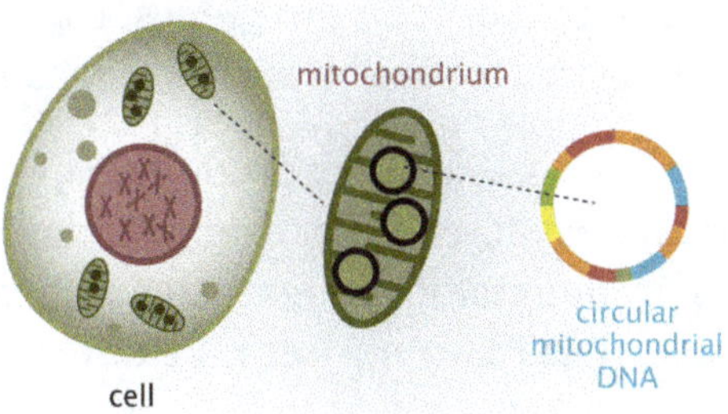

Problems with mtDNA are often difficult to diagnose. Although we know that a number of conditions occur as a result of mtDNA mutations, patients with the same disorder often display different symptoms. What is more, mitochondrial abnormalities are often responsible for multi-system disorders: for example they may affect both the muscles and the nervous system. You will remember that the mitochondria are the power plants of our cells and therefore of the whole body. For example, if a problem arises in the respiratory system, this has consequences for other systems in the body that require a lot of energy. In mitochondrial disorders it is therefore often difficult to distinguish cause from effect.

Chloroplast DNA

Chloroplast-DNA (cp-DNA) Circular DNA is like that of mitochondria but several times larger which is found in chloroplasts and other plastids and contains genes involved in the photosynthetic pathway. It is independent of nuclear DNA and is maternally inherited. Between 40 and 80 DNA molecules occur in each organelle.

Identity of Chloroplast Genes

Advances in molecular genetics have allowed scientists to take a much closer look at the chloroplast genome. The size of the genome has been determined for a number of plants and algae and ranges from 85 to 292 kilo base pairs (one kb equals one thousand base pairs), with most being between 120 kb and 160 kb. The complete DNA sequences for several different chloroplast genomes of plants and algae have been determined.

Although a simple sequence does not necessarily identify the role of each gene, it has allowed the identity of a number of genes to be determined, and it has allowed scientists to estimate the total number of genes. In terms of genome size, chloroplast genomes are relatively small and contain slightly more than one hundred genes.

Roughly half of the chloroplast genes produce either RNA molecules or polypeptides that are important for protein synthesis. Some of the RNA genes occur twice in the chloroplast genomes of almost all land plants and some groups of algae.

The products of these genes represent all the ingredients needed for chloroplasts to carry out transcription and translation of their own genes. Half of the remaining genes produce polypeptides directly required for the biochemical reactions of photosynthesis.

What is unusual about these genes is that their products represent only a portion of the polypeptides required for photosynthesis. For example, the very important enzyme ATPase, the enzyme that uses proton gradient energy to produce the important energy molecule adenosine triphosphate (ATP), comprises nine different polypeptides.

Six of these polypeptides are products of chloroplast genes, but the other three are products of nuclear genes that must be transported into the chloroplast to join with the other six polypeptides to make active ATPase.

Another notable example is the enzyme ribulose biphosphate carboxylase (RuBP carboxylase), which is composed of two polypeptides. The larger polypeptide, called rbcL is a product of a chloroplast gene, whereas the smaller polypeptide is the product of a nuclear gene.

The last thirty or so genes remain unidentified, their presence is inferred because they have DNA sequences that contain all the components found in active genes. These kinds of genes are often called "open reading frames" (ORFs) until the functions of their polypeptide products are identified.

Nuclear DNA

Nuclear DNA is the most determinative type of DNA i.e. it is more informative in DNA identification. The more concentrated the biological samples are (blood, sperm, saliva, sweat), the more chances we have to extract nuclear DNA. However nuclear DNA is more sensitive to environmental influences and therefore samples should be protected from humidity, light and contaminants.

Nuclear DNA is found within all cells (apart from red blood cells) and is organized into long structures of chromosomes. DNA is inherited from both parents and allows for the genetic identification of an individual. The technology used for genetic profiling studies specific regions (loci) of the DNA, called STRs (Short Tandem Repeats: usually tetrameric repeating sequences), that are popular due to their high level of variability that allows one to distinguish one DNA profile form another. The number of repeats of the STR sequence is determined by each chromosome (half of which is inherited from the father and half from the mother).The analysis of several regions of DNA (at least 16) determines the type of STR genetic fingerprint (gene profile) of each individual, which statistically occurs only once in 180 billion people and is now the most accepted method worldwide.

Some Facts about Nuclear DNA

1. Nuclear DNA is double stranded and linear.
2. Nuclear DNA or genome is enveloped by nuclear membrane.
3. Nuclear genome is made of 3.3 billion DNA base pairs.
4. Only two copies of nuclear genome per somatic cell.
5. Nuclear genome has 20,000-25,000 genes including mitochondrial genes.
6. Nuclear genes codes for all proteins required for its function.
7. Nuclear genome has introns or non-coding DNA and accounts for 93% of total DNA.
8. Nuclear genes show universal codon pattern.

   Examples: In nucleus, AUG codes for methionine and UGA is a stop codon.
9. Transcription of nuclear genes is monocistronic, an mRNA with sequence coding for a single protein.
10. The nuclear genome is inherited equally from both parents.

DNA Condensation

Endocytosis of plasmid DNA can be achieved if the DNA can be condensed to 100 nm or less in diameter, while access to tumor tissue from the vasculature is limited to particles less than 70 nm in diameter. It is therefore important to condense plasmid DNA to as small a size as possible to facilitate gene transfer, and indeed the size of the condensed DNA complex may be one of the most important factors for successful *in vivo* gene delivery.

Polylysine was used to condense DNA as early as 1969 and was initially used merely as a model for the interaction of biopolymers such as DNA and histone proteins. DNA is a highly negatively charge polymer due to the repeating phosphate groups along the polymer backbone. The interaction with cationic polymers such as polylysine is therefore an electrostatic one. The exact parameters that the govern the ability of cationic polymers to condense DNA are still under study, although it is generally accepted that neutralisation of the charges on the DNA molecule, followed by hydrophobic collapse as water is displaced from the DNA structure, may play an important role.

A figure showing the principle behind DNA condensation with polycations is shown below, along with some transmission electron microscopy images of DNA before and after condensation.

In 1987, Wu & Wu proposed using polylysine-condensed DNA as a gene delivery vector. Since then it has been used as the basis of most cationic polymer-based vectors. The size of the polylysine/DNA complexes varies according to the molecular weight of the polymer used and the formation conditions such as the charge ratio between the polymer and DNA, the salt concentration and the temperature. However, complex sizes of 50-300 nm diameter can generally be achieved, while careful choice of the polymer molecular weight can produce complexes with an average diameter of less than 30 nm. Careful control of mixing conditions and salt concentrations has been reported to produce complexes as small as 12 nm in diameter.

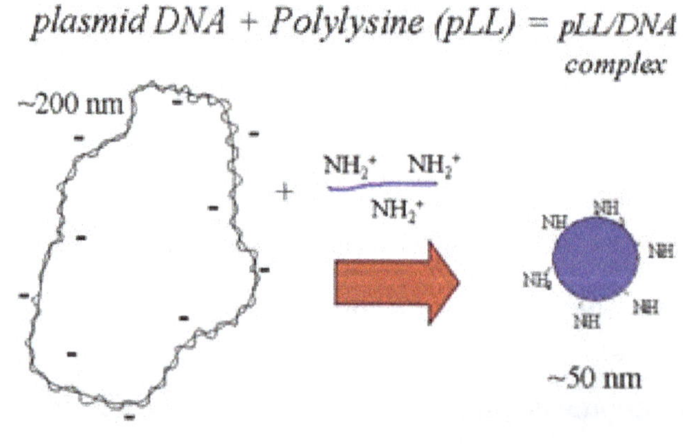

plasmid DNA + Polylysine (pLL) = pLL/DNA complex

~200 nm

NH_3^+ NH_3^+ NH_3^+

~50 nm

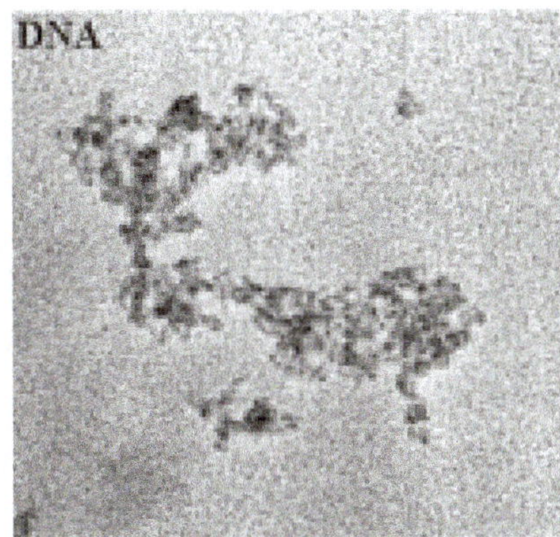

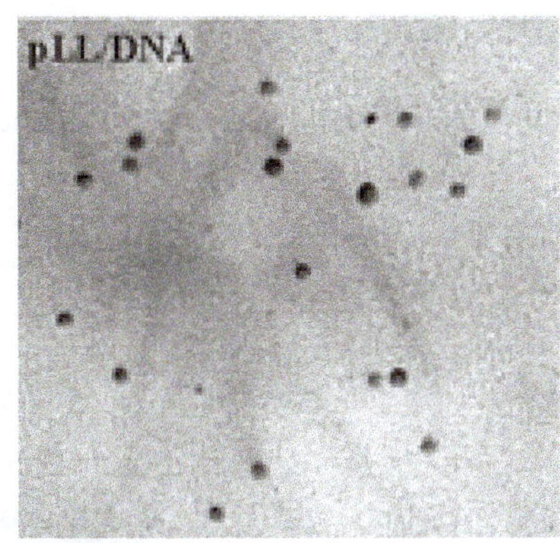

One of the major advantages of using cationic polymers to condense DNA is that very large genes can be used. Viral vectors are limited by the amount of genetic material that can be inserted into the viral genome. This limitation does not apply to polymer-based vectors, and even DNA molecules as large as 45 Kb have been successfully condensed. This permits the incorporation of gene regulatory regions such as locus control regions that may afford better control of gene expression.

Once DNA condensation has been achieved and a satisfactory complex has been produced it is possible to incorporate functional groups such as targeting moieties, nuclear localisation or targeting peptides, endosomalytic peptides andhydrophilic shielding.

DNA Fragmentation

The DNA Fragmentation is one of several elements of semen quality used to assess the male fertility potential. Sperm's ability to fertilize an egg is dependant on healthy DNA. But some sperm are fragmented. Healthy DNA is arranged in a double-helix spiral bound by cross-bonds resembling a ladder. DNA damage means the bridges become unstable or broken causing instability in the DNA ladder. That instability is referred to as fragmentation or damage of the DNA. If there is a high amount, you'll likely see a reduction in male fertility, poor embryo development and lower rates of implantation.

DNA Fragmentation Index (DFI)

DNA Fragmentation Index (DFI) is used to quantify DNA damage in the sperm. DFI is inversely related to sperm's ability to fertilize an egg and produce a viable embryo. DFI of less than 15% in sample is considered optimal. DFI between 16-29% is considered to be good or fair fertility potential. Sperm with DFI over 30% is considered to have poor fertility potential.

DNA Fragmentation Testing

DNA fragmentation testing is a type of male fertility test which measures the amount of damaged DNA in a sperm sample. All men have some amount of damage to their sperm DNA, but high percentages of damage may indicate greater difficulty achieving pregnancy with IUI and IVF.

Testing for Sperm Quality

For decades, fertility testing for men has relied almost solely on a basic semen analysis, which examines sperm concentration, morphology and motility. Still, around 30% of the 5 million men in the US who are facing infertility still do not know why they are infertile. This suggests that a semen analysis is inadequate when it comes to evaluating male fertility potential. While it can check for the presence of motile sperm that are shaped normally, it does not provide much insight on the sperm's ability to contribute to a successful pregnancy.

Sperm DNA Fragmentation

Sperm DNA fragmentation refers to the amount of damage seen in sperm DNA. Essentially, calculating what percentage of the sperm has breaks in the DNA. Sperm are very small yet they need to carry massive amounts of genetic information to the egg. With so much DNA getting packed into a tiny sperm cell, it turns out not everything arrives intact as intended.

Over the last few decades, significant research has been done to look at how the amount of damage observed in sperm DNA can be used as a marker for male infertility. While our understanding of sperm DNA damage has grown, the exact cause remains unknown. One important observation is that the amount of sperm damage detected does not always correlate with semen parameters. In other words, you can have a normal sperm count, motility, and morphology, and still have large amounts of sperm DNA damage. Interestingly, the egg has some capacity to repair damaged sperm DNA upon fertilization; however, scientists believe there may be a threshold of DNA damage beyond the repair capacity of eggs.

Sperm Chromatin Structure Analysis (SCSA) Testing

Many tests have been developed to measure sperm DNA damage, but they all use slightly different procedures and result in different standards of measurement. The SCSA test is the most widely used method. It involves treating sperm with a chemical dye; broken sperm DNA turns red, normal sperm DNA turns green.

Afterwards, the sample is run through a computerized machine with a special software which reads about 5,000 sperm cells categorizing the ratio of green-to-red sperm cell to figure out the Fragmentation Index (DFI). It is possible that SCSA testing can detect damage in some 20% of couples.

DNA Methylation

DNA methylation is an epigenetic mechanism used by cells to control gene expression. A number of mechanisms exist to control gene expression in eukaryotes, but DNA methylation is a commonly used epigenetic signaling tool that can fix genes in the "off" position.

Over recent decades, scientists have made various discoveries about DNA methylation and how vital it is to a number of cellular processes such as embryonic development, X-chromosome inactivation, genomic imprinting, gene suppression, carcinogenesis and chromosome stability. Researchers have linked abnormal DNA methylation to several adverse outcomes, including human diseases.

DNA contains combinations of four nucleotides which include cytosine, guanine, thymine and adenine. DNA methylation refers to the addition of a methyl (CH3) group to the DNA strand itself, often to the fifth carbon atom of a cytosine ring. This conversion of cytosine bases to 5-methylcytosine is catalysed by DNA methyltransferases (DNMTs). These modified cytosine residues usually lie next to a guanine base (CpG methylation) and the result is two methylated cytosines positioned diagonally to each other on opposite strands of DNA.

Different DNMTs work together either as *de novo* DNMTs, establishing the methyl group pattern on a sequence of DNA or as maintenance DNMTs that copy the methylation pattern on an existing strand of DNA to its new partner following replication. Methylation is sparse but global in mammals, found in CpG sequences across the entire genome, aside from certain stretches (of around one kilobase) where the content of CpG is high (CpG islands). When those sequences are methylated, the result can be the inappropriate silencing of genes such as tumor suppression genes.

The global distribution of methylation in mammals has posed a challenge to researchers in terms of finding out whether methylation is a default state or is targeted at specific gene sequences. However, CpG islands are generally found in close proximity to transcription start sites, suggesting there is an established recognition system.

In addition to DNA methylation being vital to healthy growth and development, it also enables the expression of retroviral genes to be suppressed, along with other potentially dangerous sequences of DNA that have entered and may damage the host.

Another important purpose of DNA methylation is the formation of the chromatin structure, which enables a single cell to grow into a complex multicellular organism made up of different tissues and organs. Scientists have established that some *de novo* DNMTs are components of chromatin-remodeling complexes that achieve remodeling by performing on the spot DNA methylation to fix in place the closed shape of chromatin.

DNA methylation plays key roles in gene expression and regulation. It is an epigenetic signaling tool that locks genes in the "off position" and is an important component in various cellular processes such as genomic imprinting, embryonic development, maintenance of chromosome stability, and X-chromosome inactivation. Scientists have linked abnormal methylation to various devastating outcomes, including human diseases.

DNA methylation involves the addition of a methyl group to the 5-carbon of the cytosine ring, which results in 5-methylcytosine or 5-mC. These methyl groups inhibit transcription by occupying the DNA's major groove. 5-mC constitutes about 1.5% of genomic DNA.

Regulation of DNA methylation by methyltransferases

DNA methylation is catalyzed by DNA methyltransferases (DNMTs) and is controlled at many different points in cellular processes. Three types of DNMTs, namely, DNMT1, DNMT3a, and DNMT3b are needed to establish and maintain DNA methylation patterns.

DNMT1 seems to be involved in the maintenance of DNA methylation patterns that are already established, while DNMT3a and DNMT3b seem to be involved in the establishment of new DNA methylation patterns. In the case of diseases such as cancer, this may be different, in that both DNMT1 and DNMT3b may be involved in maintaining hypermethylation in cancer cells.

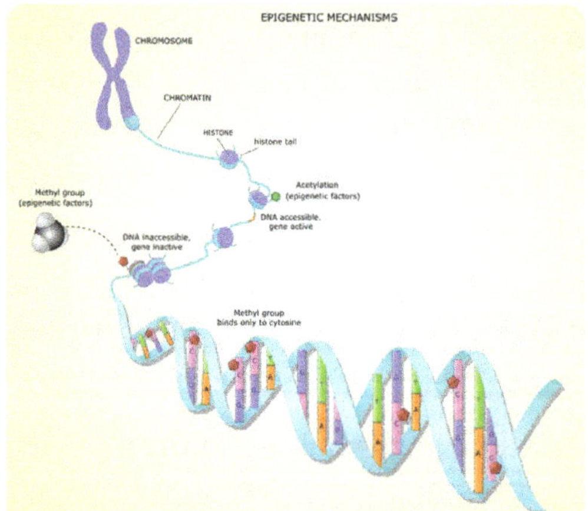

DNA methylation is a reversible epigenetic modification. The removal of a methyl group or demethylation is needed to reprogram genes and is therefore also important in mechanisms of disease such as tumor growth. Demethylation is catalyzed by enzymes called DNA demethylases.

The role of DNA methylation in gene expression varies across different kingdoms of organisms. 5'—C—phosphate—G—3' (CpG) methylation is distributed fairly globally in mammals, whereas among invertebrates, the methylation pattern is generally "mosaic," with heavily methylated DNA regions being interspersed with regions that are not methylated.

In mammals, the global methylation pattern makes it challenging to determine whether methylation is a default state or in fact targeted to specific sequences of genes. However, the CpG islands usually occur close to transcription start sites, suggesting that there is a system for recognition.

In plants, as much as 50% of cytosine residues are in a methylated state, while in fungi, only repeated sequences of DNA are methylated, with methylation completely absent in some species.

The significance of 5-mC as a key epigenetic modification in gene expression is widely recognized. For instance, a decrease in global DNA methylation or DNA hypomethylation is likely to be the result of methyl deficiency caused by various environmental factors and it has been suggested as a molecular marker in many biological processes, including cancer.

Determining the 5-mC content or global methylation in cells that have been impacted by environmental factors or become diseased, could offer information to aid disease detection and analysis. Moreover, the presence of 5-fC, the intermediate of DNA demethylation, in tissues and cells may serve as a marker for active DNA demethylation.

DNA Methylation and Disease

Since DNA methylation plays such an important role in gene expression, it seems obvious that faulty methylation could have devastating consequences, including human disease.

Researchers have conducted many studies examining the association between errors in this methylation and diseases such as cancer, muscular dystrophy, lupus, and various birth defects. This is hoped to provide valuable insights to improve the understanding, treatment and prevention of these disorders.

Studies have focused on the connection between DNA methylation and tumor suppressor genes and cancer. DNA hypermethylation has been shown to silence tumor suppressor genes in cancer cells.

By contrast, cancer cell genomes have exhibited hypomethylation overall in comparison to healthy cells, but with hypermethylation of the genes involved in tumor cell invasion, DNA repair, regulation of the cell cycle and other processes where silencing leads to metastasis.

The detection of hypermethylation in some cancers such as colon cancer is possible early on in the course of disease and may serve as a valuable biomarker for the condition.

DNA Shuffling

DNA shuffling refers to in vitro homologous recombination of pools of selected genes by random fragmentation and polymerase chain reassembly. This method is introduced by Stemmer in 1994. It involves the digestion of a gene by DNaseI into random fragments, and the reassembly of those fragments into a full-length gene by primer-less PCR: the fragments prime on each other based on sequence homology, and recombination occurs when fragments from one copy of a gene anneal to fragments from another copy, causing a template switch, or crossover event.

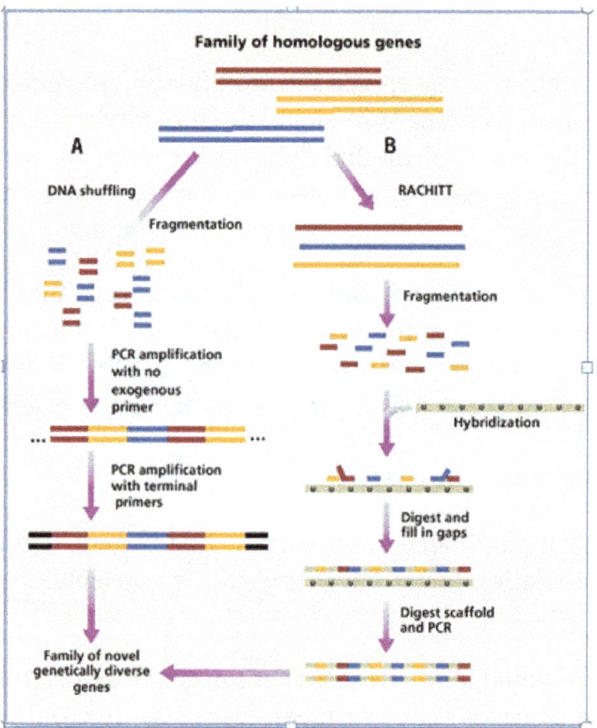

Advantages of DNA Shuffling

- Error-prone PCR can be used to mutagenize a mixture of fragments of unknown sequence. The published error-prone PCR protocols do not allow amplification of DNA fragments >0.5-1.0 kb, limiting their practical application.
- This method can be applied to sequences >1 kb.
- It has a mutagenesis rate similar to error-prone PCR.
- It also works with pools of unknown sequence.
- This method makes possible to remove neutral mutations by backcrossing with excess parental or wild-type DNA.

RNA

RNA molecules are single-stranded nucleic acids composed of nucleotides. RNA plays a major role in protein synthesis as it is involved in the transcription, decoding, and translation of the genetic code to produce proteins. RNA stands for ribonucleic acid and like DNA, RNA nucleotides contain three components:

- A Nitrogenous Base;
- A Five-Carbon Sugar;
- A Phosphate Group.

RNA nitrogenous bases include adenine (A), guanine (G), cytosine (C) and uracil (U). The five-carbon (pentose) sugar in RNA is ribose. RNA molecules are polymers of nucleotides joined to one another by covalent bonds between the phosphate of one nucleotide and the sugar of another. These linkages are called phosphodiester linkages.

Although single-stranded, RNA is not always linear. It has the ability to fold into complex three-dimensional shapes and form hairpin loops. When this occurs, the nitrogenous bases bind to one another. Adenine pairs with uracil (A-U) and guanine pairs with cytosine(G-C). Hairpin loops are commonly observed in RNA molecules such as messenger RNA (mRNA) and transfer RNA (tRNA).

Types of RNA

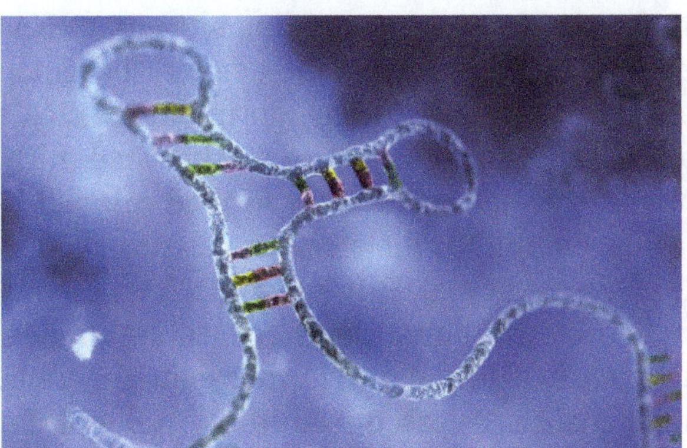

Fig: Although single stranded, RNA is not always linear. It has the ability to fold into complex three dimensional shapes and form hairpin loops.

RNA molecules are produced in the nucleus of our cells and can also be found in the cytoplasm. The three primary types of RNA molecules are messenger RNA, transfer RNA and ribosomal RNA.

- Messenger RNA (mRNA): plays an important role in the transcription of DNA. Transcription is the process in protein synthesis that involves copying the genetic information contained within DNA into an RNA message. During transcription, certain proteins called transcription factors unwind the DNA strand and allow the enzyme RNA polymerase to transcribe only a single strand of DNA. DNA contains the four nucleotide bases adenine (A), guanine (G), cytosine (C) and thymine (T) which are paired together (A-T and C-G). When RNA polymerase transcribes the DNA into a mRNA molecule, adenine pairs with uracil and cytosine pairs with guanine (A-U and C-G). At the end of transcription, mRNA is transported to the cytoplasm for the completion of protein synthesis.

- Transfer RNA (tRNA): plays an important role in the translation portion of protein synthesis. Its job is to translate the message within the nucleotide sequences of mRNA into specific amino acid sequences. The amino acid sequences are joined together to form a protein. Transfer RNA is shaped like a clover leaf with three hairpin loops. It contains an amino acid attachment site on one end and a special section in the middle loop called the anticodon site. The anticodon recognizes a specific area on mRNA called a codon. A codon consists of three continuous nucleotide bases that code for an amino acid or signal the end of translation. Transfer RNA along with ribosomes read the mRNA codons and produce a polypeptide chain. The polypeptide chain undergoes several modifications before becoming a fully functioning protein.
- Ribosomal RNA (rRNA): is a component of cell organelles called ribosomes. A ribosome consists of ribosomal proteins and rRNA. Ribosomes are typically composed of two subunits: a large subunit and a small subunit. Ribosomal subunits are synthesized in the nucleus by the nucleolus. Ribosomes contain a binding site for mRNA and two binding sites for tRNA located in the large ribosomal subunit. During translation, a small ribosomal subunit attaches to a mRNA molecule. At the same time, an initiator tRNA molecule recognizes and binds to a specific codon sequence on the same mRNA molecule. A large ribosomal subunit then joins the newly formed complex. Both ribosomal subunits travel along the mRNA molecule translating the codons on mRNA into a polypeptide chain as they go. Ribosomal RNA is responsible for creating the peptide bonds between the amino acids in the polypeptide chain. When a termination codon is reached on the mRNA molecule, the translation process ends. The polypeptide chain is released from the tRNA molecule and the ribosome splits back into large and small subunits.

MicroRNAs

Some RNAs, known as small regulatory RNAs, have the ability to regulate gene expression. MicroRNAs (miRNAs) are a type of regulatory RNA that can inhibit gene expression by halting translation. They do so by binding to a specific location on mRNA, preventing the molecule from being translated. MicroRNAs have also been linked to the development of some types of cancers and a particular chromosome mutation called a translocation.

Transfer RNA

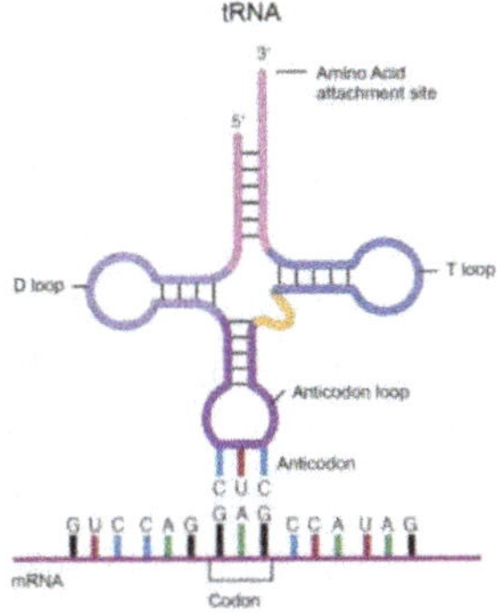

Transfer RNA (tRNA) is an RNA molecule that assists in protein synthesis. Its unique shape contains an amino acid attachment site on one end of the molecule and an anticodon region on the opposite end of the amino acid attachment site. During translation, the anticodon region of tRNA recognizes a specific area on messenger RNA (mRNA) called a codon. A codon consists of three continuous nucleotide bases that specify a particular amino acid or signal the end of translation. The tRNA molecule forms base pairs with its complementary codon sequence on the mRNA molecule. The attached amino acid on the tRNA molecule is therefore placed in its proper position in the growing protein chain.

RNA Interference

RNA interference (RNAi) is a regulatory system occurring within eukaryotic cells (cells with a clearly defined nucleus) that controls the activity of genes. RNAi functions specifically to silence, or deactivate, genes.

The ability of interfering RNA to silence genes was discovered in the 1990s by American scientists Andrew Z. Fire and Craig C. Mello, who shared the 2006 Nobel Prize for Physiology or Medicine for their work. Fire and Mello successfully inhibited the expression of specific genes by introducing short double-stranded RNA (dsRNA) segments into the cells of nematodes (*Caenorhabditis elegans*). The dsRNA segments underwent enzymatic processing that enabled them to attach to molecules of messenger RNA(mRNA) possessing complementary nucleotide sequences. The attachment of the two RNAs inhibited the translation of the mRNA molecules into proteins.

RNAi in Nature

Although Fire and Mello's work involved the experimental introduction of interfering RNA into cells, gene silencing by RNAi is a natural genetic mechanism in eukaryotes that takes place following transcription (the synthesis of mRNA from DNA). Special microRNA(miRNA) segments, each of which is approximately 20 nucleotides in length, are encoded by the genomes of eukaryotic organisms. Each miRNA is produced from a precursor transcript (pre-miRNA). After the pre-miRNA migrates from the nucleus into the cytoplasm, it is cleaved into a mature miRNA by an enzyme known as DICER. The mature miRNA molecule then binds to an RNA-induced silencing complex (RISC), which contains multiple proteins, including a ribonuclease enzyme. The miRNA nucleotide sequence directs the protein complex to bind to a complementary sequence of mRNA. Once bound to the mRNA, the miRNA-RISC complex then enzymatically cleaves targeted sites on the mRNA molecule, thereby inhibiting the translation of the gene into a protein, which effectively silences the gene.

RNAi plays an important role not only in regulating genes but also in mediating cellular defense against infection by RNA viruses, including influenza viruses and rhabdoviruses, a group that contains the causative agent of rabies. In fact, a number of plants and animals have evolved antiviral RNAi genes that encode short segments of RNA molecules with sequences that are complementary to viral sequences. This complementarity enables interfering RNA produced by the cell to bind to and inactivate specific RNA viruses.

RNAi also is an innate mechanism by which cells can suppress the activity of transposons, or “jumping genes.” Certain types of transposable elements are able to produce mobile copies of them-selves, which subsequently are inserted into various regions of the genome, giving rise to repetitive sequences of DNA. These insertions generally are of little concern. However, some insertions lead to increased or decreased gene activity and can give rise to disease in humans. For example, certain types of cancer and Duchenne muscular dystrophy, a hereditary muscle-wasting disorder, are associated with insertions of transposons.

RNAi in Research and Medicine

The discovery that genes can be silenced by segments of dsRNA that are introduced into cells in tissue culture revolutionized the study of gene function. Gene silencing by dsRNA makes use of the naturally occurring cell machinery that is involved in the processing of miRNA in eukaryotic cells. For example, each dsRNA is cleaved into small pieces by the DICER enzyme. These pieces are called short interfering RNAs (siRNAs) and are about 20 to 25 nucleotides in length. Similar to miRNA, siRNA binds to RISC and cleaves targeted sequences of mRNA.

There are different types of synthetic dsRNAs that can be employed to disrupt gene function. Commonly used molecules include siRNA, which bypasses DICER cleavage, and small hairpin RNA (shRNA), which actually is one RNA strand containing two unique siRNA segments that is folded into a double strand, with the adjacent nucleotides joined through heating (annealing) rather than complementary base pairing. This creates a structure that resembles a hairpin because it has a tight loop at one end. Inside a cell, the shRNA is cleaved into its two component siRNAs by DICER.

RNAi is an exceptionally powerful research tool. Synthetic dsRNAs are designed to prevent the expression of specific genes, thereby enabling geneticists to manipulate the activity of genes in order to better understand their functions. In addition, abnormally overactive genes contribute to certain human diseases, and silencing this activity using RNAi has become an important area of investigation. RNAi is being explored as a form of treatment for a variety of diseases, including macular degeneration, hepatitis, AIDS, Huntington disease, and cancer.

In macular degeneration, RNA sequences that block the production of a protein called vascular endothelial growth factor (VEGF) in cells of the retina can inhibit the excess growth of retinal blood vessels, which leak and lead to vision loss. RNAi treatments for macular degeneration involve the injection of "naked RNA" into the eye. The term *naked RNA* is used to distinguish this approach from those that employ viral vectors to introduce dsRNA into diseased cells. Interfering RNAs incorporated into vectors are being studied for their effectiveness in slowing tumor growth. For example, mRNA transcripts of genes known to be overactive in certain forms of cancer serve as useful targets for RNAi-based treatments, which can silence overactive genes and slow disease progression.

Factors such as ensuring that interfering RNAs reach the cells and that the viral vectors themselves do not give rise to dangerous side effects have complicated the development of RNAi therapies. Furthermore, sequence similarities between genes can result in the binding of dsRNAs to otherwise properly functioning genes. This can result in the silencing of healthy genes vital to normal cell function. Nevertheless, the technique remains promising for applications in medicine.

RNA Processing

RNA processing refers to any modification made to RNA between its transcription and its final function in the cell. These processing steps include the removal of extra sections of RNA, specific modifications of RNA bases, and modifications of the ends of the RNA.

Types of RNA Processing

There are three main types of RNA processing events: trimming one or both of the ends of the primary transcript to the mature RNA length; removing internal RNA sequences by a process called RNA splicing; and modifying RNA nucleotides either at the ends of an RNA or within the body of the RNA. We will briefly examine each of these and then discuss how they are applied to the various types of cellular RNAs.

Almost all RNAs have extra sequences at one or both ends of the primary transcripts that must be removed. The removal of individual nucleotides from the ends of the RNA strand is carried out by any of several ribonucleases (enzymes that cut RNA), called exoribonucleases. An entire section of RNA sequence can be removed by cleavage in the middle of an RNA strand. The enzymes responsible for the cleavage in this location are called endoribonucleases. Each of these ribonucleases is targeted so that it only cleaves particular RNAs at particular places.

RNA splicing is similar to trimming in that it removes extra RNA sequences, but it is different because the sequence is removed from the middle of an RNA and the two flanking pieces are joined together again (see figure). The part of the RNA that is removed is called an intron, whereas the two pieces that are joined together, or spliced, are called exons. Just as with the cleavage enzymes, the splicing machinery recognizes particular sites within the RNA, in this case the junctions between exons and introns, and cleaves and rejoins the RNA at those positions.

Modification of RNA nucleotides can occur at the ends of an RNA molecule or at internal positions. Modification of the ends can protect the RNA from degradation by exoribonucleases and can also act as a signal to guide the transport of the molecule to a particular subcellular compartment. Some internal modifications, particularly of tRNAs and rRNAs, are necessary for these RNAs to carry out their functions in protein synthesis. Some internal modifications of mRNAs change the sequence of the message and so change the amino acid sequence of the protein coded for by the mRNA. This process is called RNA editing. As with the other types of RNA processing, the enzymes that modify RNAs are directed to specific sites on the RNA.

Processing of Various Classes of RNAs

Ribosomal RNAs are synthesized as long primary transcripts that contain several different rRNAs separated by spacer regions (see figure). The individual rRNAs are cut apart by endoribonucleases that cleave within the spacer regions. Other enzymes then trim the ends to their final length. Ribosomal RNAs are also modified at many specific sites within the RNA. Ribosomal RNA synthesis and processing occurs in a special structure within the nucleus called the nucleolus. The mature rRNAs bind to ribosomal proteins within the nucleolus and the assembled ribosomes are then transported to the cytoplasm to carry out protein synthesis.

Transfer RNAs are transcribed individually from tRNA genes. The primary transcripts are trimmed at both the 5′ and 3′ ("five prime," or "upstream" and "three prime," or "downstream") ends, and several modifications are made to internal bases. Many eukaryotic tRNAs also contain an intron, which must be removed by RNA splicing. The finished tRNAs are then transported from the nucleus to the cytoplasm.

Messenger RNAs are transcribed individually from their genes as very long primary transcripts. This is because most eukaryotic genes are divided into many exons separated by introns. Genes may contain from zero to more than sixty introns, with a typical gene having around ten. Introns are spliced out of primary RNA transcripts by a large structure called the spliceosome. The spliceosome does not move along the RNA but is assembled around each intron where it cuts and joins the RNA to remove the intron and connect the exons. This must be done many times on a typical primary transcript to produce the mature mRNA.

In addition to removal of the introns, the mRNA is modified at the 5′ end by the addition of a special "cap" structure that is later recognized by the translation machinery. The mRNA is also trimmed at the 3′end and several hundred adenosine nucleotides are added. This modification, which is called either polyadenylation or poly (A) addition, helps stabilize the 3′ end against degradation and is also recognized by the translation machinery. Finally, the processed mature mRNA is transported from the nucleus to the cytoplasm.

Some RNAs, called small nuclear RNAs (snRNAs) and small nucleolar RNAs (snoRNAs), are processed in the nucleus and are themselves part of the RNA processing systems in the nucleus. Most snRNAs are involved in mRNA splicing, while most snoRNAs are involved in rRNA cleavage and modification.

RNA Processing and the Human Genome

The fact that most human genes are composed of many exons has some important consequences for the expression of genetic information.

First, we now know that many genes are spliced in more than one way, a phenomenon known as alternative splicing. For example, some types of cells might leave out an exon from the final mRNA that is left in by other types of cells, giving it a slightly different function. This means that a single gene can code for more than one protein. Some complicated genes appear to be spliced to give hundreds of alternative forms. Alternative splicing, therefore, can increase the coding capacity of the genome without increasing the number of genes.

A second consequence of the exon/intron gene structure is that many human gene mutations affect the splicing pattern of that gene. For example, a mutation in the sequence at an intron/exon junction that is recognized by the spliceosome can cause the junction to be ignored. This causes splicing to occur to the next exon in line, leaving out the exon next to the mutation. This is called exon skipping and it usually results in an mRNA that code for a nonfunctional protein. Exon skipping and other errors in splicing are seen in many human genetic diseases.

Nucleotides

A nucleotide is an organic molecule that is the building block of DNA and RNA. They also have functions related to cell signaling, metabolism, and enzyme reactions. A nucleotide is made up of three parts: a phosphate group, a 5-carbon sugar, and a nitrogenous base. The four nitrogenous bases in DNA are adenine, cytosine, guanine, and thymine. RNA contains uracil, instead of thymine. A nucleotide within a chain makes up the genetic material of all known living things. They also serve a number of function outside of genetic information storage, as messengers and energy moving molecules.

A series of three nucleotides within the DNA is known as a *codon*, and directs the proteins within the cell to attach a specific protein to a series specified by the rest of the DNA. Special codons even specify to the machinery where to stop and start the process. *DNA translation*, as it is known, converts the information from DNA into the language of proteins. This chain of amino acids can be properly folded, and provide one of many functions within the cell.

Nucleotide Structure

Nucleotide structure is simple, but the structure they can form together is complex. Below is an image of DNA. This molecule consists of two strands which wrap around each other, forming *hydrogen bonds* in the middle of the structure for support. Each nucleotide within has a specific structure which enables this formation.

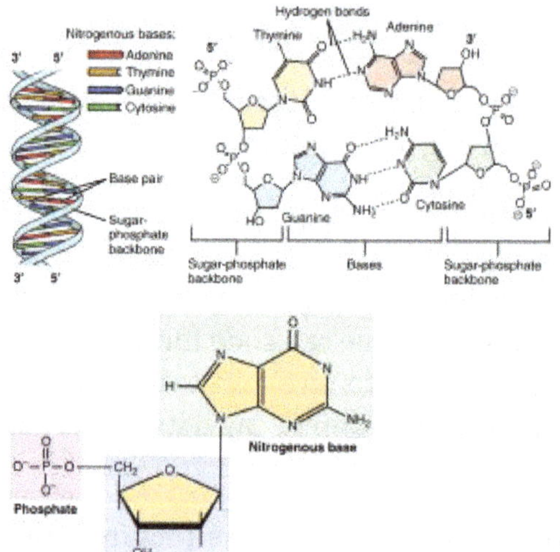

Nitrogenous Base

The nitrogenous base is the central information carrying part of the nucleotide structure. These molecules, which have different exposed functional groups, have differing abilities to interact with each other. As in the image, the idea arrangement is the maximum amount of hydrogen bonds between nucleotides involved. Because of the structure of the nucleotide, only a certain nucleotide can interact with other. The image above shows thymine bonding to adenine, and guanine bonding to cytosine. This is the proper and typical arrangement.

This even formation causes a twist in the structure, and is smooth if there are no errors. One of the ways proteins are able to repair damaged DNA is that they can bind to uneven spots within the structure. Uneven spots are created when hydrogen bonding does not occur between the opposing nucleotide molecules. The protein will cut out one nucleotide, and replace it with another. The duplicate nature of the genetic strands ensures that errors like this can be corrected with a high degree of accuracy.

Sugar

The second portion of the nucleotide is the sugar. Regardless of the nucleotide, the sugar is always the same. The difference is between DNA and RNA. In DNA, the 5-carbon sugar is deoxyribose while in RNA the 5-carbon sugar is ribose. This gives genetic molecules their names; the full name of DNA is deoxyribonucleic acid, and RNA is ribonucleic acid.

The sugar, with its exposed oxygen, can bond with the phosphate group of the next molecule. They then form a bond, which becomes the *sugar-phosphate backbone.* This structure adds rigidity to the structure, as the *covalent* bonds they form are much stronger than the hydrogen bonds between the two strands. When proteins come to process and *transpose* the DNA, they do so by separating the strands and reading only one side. When they pass on, the strands of genetic material comes back together, driven by the attraction between the opposing nucleotide bases. The sugar-phosphate backbone stays connected the whole time.

Phosphate Group

The last part of nucleotide structure, the phosphate group, is probably familiar from another important molecule *ATP*. Adenosine triphosphate, or ATP, is the energy molecule that most life on Earth relies upon to store and transfer energy between reactions. ATP contains three phosphate groups, which can store a lot of energy in their bonds. Unlike ATP, the bonds formed within a nucleotide are known as *phosphordiester bonds*, because they happen between the phosphate group and the sugar molecule.

During DNA replication, an enzyme known as *DNA polymerase* assembles the correct nucleotide bases, and begins organizing them against the chain it is reading. Another protein, *DNA ligase*, finished the job by creating the phosphodiester bond between the sugar molecule of one base and the phosphate group of the next. This creates the backbone of a new genetic molecule, able to be passed to the next generation. DNA and RNA contain all the genetic information necessary for cells to function.

Nucleotide Function

Besides being the basic unit of genetic material for all living things, a nucleotide can have other functions as well. A nucleotide can be a base in another molecule, such as adenosine triphosphate (ATP), which is the main energy molecule of the cell. They are also found in coenzymes like NAD and NADP, which come from ADP; these molecules are used in many chemical reactions that play roles in metabolism. Another molecule that contains a nucleotide is cyclic AMP (cAMP), a messenger molecule that is important in many processes including the regulation of metabolism and transporting chemical signals to cells. Nucleotides not only make up the building blocks of life but also form many different molecules that function to make life possible.

Polynucleotide

Polynucleotide molecules are a biopolymer consisting of an elongated, linear sequence of nucleotides that are linked together by the ester linkages between the hydroxyl group of the sugar component of the next nucleotide and the phosphoryl group of nucleotide.

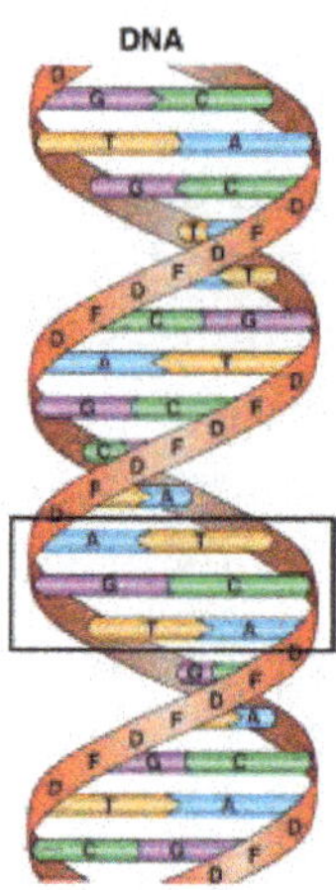

The DNA called a polynucleotide because the DNA molecule is composed of nucleotides- deoxyadenylate (A), deoxyguanylate (G), deoxycytidylate (C), and deoxythymidylate (thymidylate) (T) which are combined together to create long chains called a polynucleotide. As per the DNA structure, the DNA consists of two chains of the polynucleotide, each in the form of a spherical spiral.

Nucleobase

A nucleobase is one of the molecules of a nucleotide that carries information. It is also a component of DNA and RNA. The helix strands of DNA are connected between themselves by pairs of nucleobases. The five principal nucleobases found in DNA and RNA are adenine, cytosine, guanine, thymine and uracil (present only in RNA). The nucleus of the nucleobase is called purine.

Adenine

Adenine is one of the nucleobases present in deoxyribonucleic acid (DNA) and ribonucleic acid (RNA), the genetic information stored within organisms. It is a substance often studied in biochemistry because of its many important roles in the bodies of organisms. It has the chemical formula $C_5H_5N_5$. It is a purine, meaning that it is a kind of organic compound that is composed of carbon and nitrogen atoms arranged in the form of two rings.

DNA and RNA are extraordinarily important nucleic acids because they contain the genetic information used for the growth, repair, development, and reproduction of all organisms. They are each made up of four nucleobases: DNA is composed of adenine, thymine, guanine, and cytosine; RNA is composed of the same, but with uracil instead of thymine. The arrangements of these nucleobases determine the exact nature of the genetic code contained in the DNA or RNA. Adenine is one of these nucleobases so it is of the utmost importance to the genetic structure of all living organisms.

In DNA, adenine bonds only to thymine. It does so with two strong hydrogen bonds, so the bond is difficult to break and the code is difficult to damage. In RNA, adenine bonds with uracil; the particular kinds of reactions that RNA is involved in favor uracil to thymine. In both cases, the particular arrangement of nucleobases determines the genetic properties of the nucleic acid.

It was initially thought that Adenine was actually vitamin B_4. It is not considered to be a direct part of the B vitamin family anymore, though some B vitamins do bind with it with varying effects. This is most notably true of niacin and riboflavin, which bind with it to form cofactors, which are required for some proteins to function properly.

Adenine is not exclusively found in nucleic acids; many different substances, such as some blends of tea, actually contain the nucleobase. It can also form a variety of compounds that are very common in nature and in some foods and drinks. Cobalamim more commonly referred to as vitamin B_{12}, is actually a compound of adenine known for its energizing effects and is a natural antidepressant. Adenosine triphosphate (ATP) is another compound that contains adenine; it is known for its role as a major energy source that is derived from cellular respiration. Glucose is broken down into ATP, which is a very significant energy-containing molecule used by a vast variety of organisms.

Cytosine

Cytosine is one of the five main nucleobases used in storing and transporting genetic information within a cell in the nucleic acids DNA and RNA. The other four nucleobases are adenine, guanine, thymine and uracil. Cytosine, thymine, and uracil are pyrimidine derivatives, while guanine and adenine are purine derivatives. The nucleoside of cytosine is cytidine.

In DNA, cytosine (C) and thymine (T) form hydrogen bonds with their complementary purine derivatives, guanine (G) and adenine (A). In RNA, the complement of adenine is uracil (U) instead of thymine. Thus, cytosine, along with adenine and guanine, is present in both DNA and RNA, whereas thymine is usually seen only in DNA and uracil only in RNA.

In Watson-Crick base pairing, cytosine forms three hydrogen bonds with guanine. From the point of view of structure, it is remarkable that cytosine, with its three binding sites, only attaches to guanine in DNA, while adenine, with two sites for hydrogen binding, only attaches to thymine. The way these hydrogen bonds hold the strands of the nucleic acid together to form the double helix, yet allowing the strands to "unzip" for replication and transcription, is simply amazing from a design point of view.

Cytosine can also be a part of a nucleotide other than related to DNA or RNA. As cytidine triphosphate (CTP), it can act as a co-factor to enzymes, and can transfer a phosphate to convert adenosine diphosphate (ADP) to adenosine triphosphate (ATP).

Properties

Cytosine is a pyrimidine derivative, with a heterocyclic, aromatic ring, and two substituents attached (an amine group at position four and a keto group at position two). *Heterocyclic* compounds are organic compounds (those containing carbon) that contain a ring structure containing atoms in addition to carbon—such as sulfur, oxygen, or nitrogen—as part of the ring. *Aromaticity* is a chemical property in which a conjugated ring of unsaturated bonds, lone pairs, or empty orbitals exhibit stabilization stronger than would be expected by the stabilization of conjugation alone. In organic chemistry, a *substituent* is an atom or group of atoms substituted in place of a hydrogen atom on the parent chain of a hydrocarbon.

In DNA and RNA, cytosine is paired with guanine. However, it is inherently unstable, and can change into uracil (spontaneous deamination). This can lead to a point mutation if not repaired by the DNA repair enzymes such as uracil glycosylase which cleaves a uracil in DNA.

Cytosine can also be methylated into 5-methylcytosine by an enzyme called DNA methyltransferase.

Guanine

Guanine, a two-ring molecular structure, is one of the five defining components or nucleobases found in the nucleic acids DNA and RNA; the others being adenine, cytosine, thymine, and uracil. Guanine and adenine are derived from the two-ring parent molecule purine and cytosine, thymine, and uracil are derived from the one-ring parent molecule pyrimidine.

Guanine ($C_5H_5N_5O$), comprises a six-carbon pyrimidine ring fused with a five-carbon imidazole ring to form a system stabilized by conjugated double bonds (the positions of the double bonds shift around the ring). Being unsaturated, the bicyclic molecule is planar. The guanine nucleoside (guanine bonded with a five-carbon sugar) is called guano sine and lacks only a phosphate to form a nucleotide.

In DNA, guanine and adenine form hydrogen bonds with their complementary pyrimidine derivatives, cytosine and thymine. In RNA, the complement of adenine is uracil instead of thymine. Thus, guanine, along with adenine and cytosine, is present in both DNA and RNA, whereas thymine is usually seen only in DNA and uracil only in RNA.

The ubiquitousness of guanine, which plays a central role in the DNA of all living organisms and even in RNA viruses is evidence of the connectedness and unity of all life.

Basic Properties

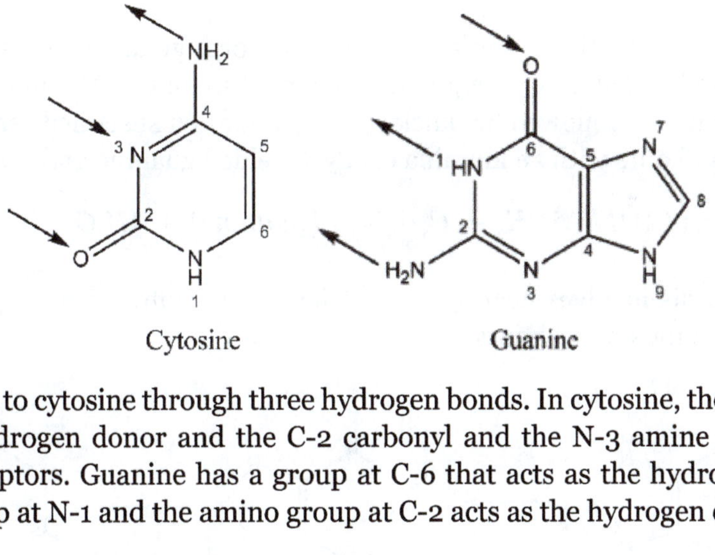

Cytosine Guanine

Guanine binds to cytosine through three hydrogen bonds. In cytosine, the amino group acts as the hydrogen donor and the C-2 carbonyl and the N-3 amine as the hydrogen-bond acceptors. Guanine has a group at C-6 that acts as the hydrogen acceptor, while the group at N-1 and the amino group at C-2 acts as the hydrogen donors.

Guanine has two tautomeric forms: the keto form (characterized by an attached OH group) and the enol form (characterized by an attached CH2 group).

Guanine can be hydrolyzed with strong acid at 180°C to glycine, ammonia, carbon dioxide, and carbon monoxide. Guanine oxidizes more readily than adenine, the other purine-derivative base in DNA and RNA. Its high melting point of 350°C reflects the strong intermolecular hydrogen bonding between the oxo and amino groups in the molecules in the crystal. Because of this intermolecular bonding, guanine is relatively insoluble in water, although it is soluble in dilute acids and bases.

The first isolation of guanine was reported in 1844 from sea bird excreta, which is known as guano and was used as a source of fertilizer. About fifty years later, Fischer determined guanine's structure and showed that uric acid can be converted to guanine. The first complete synthesis of guanine was done by Traube and remains among the best large-scale preparations.

Synthesis

Trace amounts of guanine form by the polymerization of ammonium cyanide (NH_4CN). Two experiments conducted by Levy et al., showed that heating ten mole NH_4CN at 80°C for 24 hours gave a yield of 0.0007 percent while using 0.1 mole NH_4CN frozen at -20°C for 25 years gave a 0.0035 percent yield (Levy et al. 1999). These results indicate guanine could arise in frozen regions of the primitive earth. In 1984, Yuasa reported a 0.00017 percent yield of guanine after the electrical discharge of NH_3, CH_4, C_2H_6, and 50 mL of water, followed by a subsequent acid hydrolysis. However, it is unknown if the presence of guanine was not simply a contaminant of the reaction.

$$5NH_3 + CH_4 + 2C_2H_6 + H_2O \rightarrow C_5H_8N_5O\ (guanine) + (25/2)H_2$$

A Fischer- Tropsch synthesis can also be used to form guanine, along with adenine, uracil, and thymine. Heating an equimolar gas mixture of CO, H_2, and NH_3 to 700 °C for 0.24 to 0.4 hours, followed by quick cooling, and then sustained reheating to 100-200°C for 16-44 hours with an alumina catalyst yielded guanine and uracil:

$$5CO + (1/2)H_2 + 5NH_3 \rightarrow C_5H_8N_5O\ (guanine) + 4H_2O$$

Traube's synthesis involves heating 2,4,5, triamino-1,6-dihydro-6-oxypyrimidine the with formic acid for several hours.

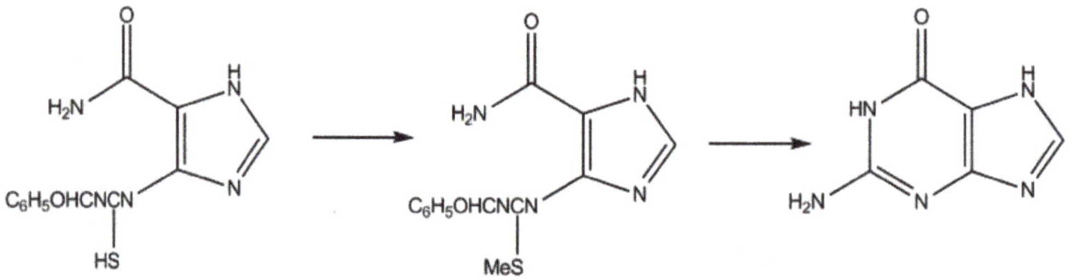

Commercial Uses

In 1656 in Paris, François Jaquin (a rosary maker) extracted from scales of some fishes the so called "pearl essence"—crystalline guanine forming G-quadruplexes. Guanine crystals are rhombic platelets composed of multiple, transparent layers but they have a high index of refraction that partially reflects and transmits light from layer to layer, thus producing a pearly luster. In the cosmetics industry, crystalline guanine is used as an additive to various products (e.g., shampoos), where it provides the pearly iridescent effect. It is also used in metallic paints and simulated pearls and plastics. Crystalline guanine provides shimmering luster to eye shadow and nail polish. It can be applied by spray, painting, or dipping, but it may irritate eyes. Alternatives include mica, synthetic pearl, aluminium and bronze particles.

Thymine

Thymine is one of the five nitrogenous nucleobases that form the basic building blocks of deoxyribonucleic acid (DNA).

Also known as *5-methyluracil*, thymine (T) is a *pyrimidine* nucleobase, which pairs with adenine (A), a *purine* nucleobase. They are joined together as a *base pair* by two hydrogen bonds, which stabilize the nucleic acid structures in DNA. When stacked with the other base pair, *guanine* (G) and *cytosine* (C), the helical structure of DNA (or RNA) is formed.

In the structure of *RNA*, thymine is replaced by the *uracil* nucleobase. As suggested by its alternative name (5-methyluracil), thymine can be derived by *methylation* of uracil at the 5th carbon (a methyl branch $-CH_3$ is added to the pyrimidine ring).

The combination of thymine, with the *pentose sugar*, *deoxyribose*, forms the *nucleoside*, deoxythymidine (alternatively named *'thymidine'*). A nucleoside is a *nucleotide* without a *phosphate group*.

Mutation and Cancer

When exposed to *ultraviolet radiation* such as sunlight, *covalent bonds* are formed between adjacent thymine molecules on the same strand of DNA, creating *thymine dimers*. This process causes damage, by causing the DNA to form 'kinks'. This inhibits the normal function of the DNA, which cannot then be replicated or *transcribed.*

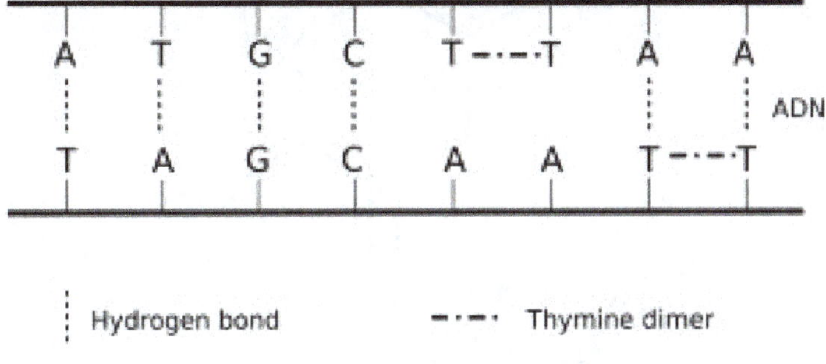

Thymine Dimers

Fortunately, most cells are able to repair damaged DNA. This can be achieved in two ways: repair enzymes called *photolyase* can break the covalent bond, using light as an energy-source for bond *cleavage*. This process is called *photoreactivation* and is possible in most organisms, although not in *placental mammals*.

The second mechanism involves an *excision enzyme*, which removes the damaged section from a single strand of DNA. The excised nucleotides are then replaced by *DNA polymerase* and a final *phosphodiester bond* (the stabilizing structure of nucleic acids) is formed by *DNA ligase*.

Thymine Structure

The formula of thymine is $C_5H_6N_2O_2$.

Properties

It is a heterocyclic, aromatic, organic compound.

Heterocyclic compounds or 'ring structures' are *cyclic compounds* (the atoms in the compound are connected to form a ring), that have atoms of at least two different elements.

An 'organic' compound contains carbon, so a heterocyclic organic compound contains atoms of carbon and one or more additional elements such as sulphur, nitrogen or oxygen.

The term *aromatic* describes a molecule which is cyclic and planar (flat), with a ring of *resonance bonds*, which give the molecule added stability. This means that it does not break apart or react with other substances easily.

Atomic Structure

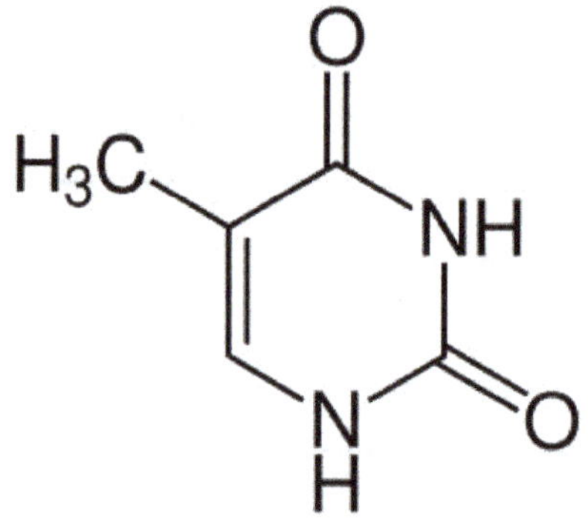

Thymin structure

In the same way as the other pyrimidines, cytosine and uracil, thymine has nitrogen (N) at the 1' and 3'; each one is bonded to hydrogen (H). The N molecule at the 3' prime forms the *glycosidic bond* – the covalent bond which joins to the deoxyribose sugar to form the nucleoside, thymidine.

At the 2' and 6' positions are *carbonyl groups* and at the left hand side is a *double bond.*

Thymine has a methyl group at the 5' position, distinguishing it from uracil, which has hydrogen at that position.

Uracil

Uracil is one of the five main nucleobases found in the nucleic acids DNA and RNA. The others are adenine, cytosine, guanine, and thymine. However, while the other four are found in DNA, Uracil is usually only found in RNA. Uracil (U), thymine (T), and cytosine (C) are pyrimidine derivatives, and guanine (G) and adenine (A) are purine derivatives.

In DNA, thymine and cytosine form hydrogen bonds with their complementary purine derivatives, adenine and cytosine, respectively. In RNA, uracil replaces thymine as the usual complement of adenine. Thus, thymine is usually seen only in DNA and uracil only in RNA. Methylation of uracil produces thymine, providing a mechanism for repair and protection of DNA and improvement of DNA replication. One of the common mutations of DNA involves two adjacent thymines.

Uracil is common and naturally occurring (Garrett and Grisham, 1997). Uracil was originally discovered in 1900 and it was isolated by hydrolysis of yeast nuclein that was found in bovine thymus and spleen, herring sperm, and wheat germ. Uracil is used in cells to carry out the synthesis of important enzymes and the biosynthesis of polysaccharides. Derivatives of uracil, developed through human creativity, include compounds that can be used as a pharmaceutical drug with anticancer properties, and those used in agriculture as herbicide and pesticide.

Properties

As a pyrimidine nucleobase, uracil is a heterocyclic aromatic organic compound. *Heterocyclic* compounds are organic compounds (those containing carbon) that contain a ring structure containing atoms in addition to carbon, such as sulfur, oxygen, or nitrogen, as part of the ring. *Aromaticity* is a chemical property in which a conjugated ring of unsaturated bonds, lone pairs, or empty orbitals exhibit stabilization stronger than would be expected by the stabilization of conjugation alone.

Uracil is a planar, unsaturated compound that has the ability to absorb light, Found in RNA, it base pairs with adenine through hydrogen bonding and is replaced by thymine in DNA. Uracil can base pair with any of the bases depending on how the molecule arranges itself on the helix, but readily pairs with adenine because the methyl group is repelled into a fixed position. Uracil is the hydrogen bond acceptor and can form up to three hydrogen bonds. Uracil can also bind with a ribose sugar to form a ribonucleoside, uridine. When a phosphate attaches to uridine, uridine 5'-monophosphate is produced.

Uracil undergoes keto-enol tautomeric shifts because of its resonance structures due to the NH_2 substituents and OH substitutents. In organic chemistry, a *substituent* is an atom or group of atoms substituted in place of a hydrogen atom on the parent chain of a hydrocarbon. Any nuclear instability the molecule may have from the lack of formal aromaticity is compensated by the cyclic-amidic stability. The keto tautomer is referred to as the lactam structure, while the enol tautomer is referred to as the lactim structure. These tautomeric forms are predominant at pH of 7. The lactam structure is the most common form of uracil.

Uracil also recycles itself to form nucleotides by undergoing a series of phophoribosyltransferase reactions. Degradation of uracil produces substrates, aspartate, carbon dioxide, and ammonia.

$$C_4H_4N_2O_2 \rightarrow H_3NCH_2CH_2COO^- + NH_4 + CO_2$$

Oxidative degradation of uracil produces urea and maleic acid in the presence of H_2O_2]] and Fe^{2+} or in the presence of diatomic oxygen and Fe^{2+}.

Uracil is a weak acid. The first site of ionization of uracil is not known. The negative charge is placed on the oxygen anion and produces a pK_a of less than or equal to 12. The basic $pK_a = -3.4$, while the acidic $pK_a = 9.38_9$. In the gas phase, uracil has four sites that are more acidic than water.

Synthesis

There are many laboratory syntheses of uracil that can be performed.

The simplest of the methods of synthesis is to add water to cytosine to produce uracil and ammonia. The most common way to synthesize uracil is by the condensation of maleic acid with urea in fuming sulfuric acid, which is also depicted below. Uracil can also be synthesized by a double decomposition of thiouracil in aqueous chloroacetic acid.

$$C_4H_5N_3O + H_2O \rightarrow C_4H_4N_2O_2 + NH_3$$
$$C_4H_4O_4 + CH_4N_2O \rightarrow C_4H_4N_2O_2 + 2\,H_2O + CO$$

Photodehydrogenation of 5,6 diuracil, which is synthesized by beta-alanine reacting with urea, produces uracil.

Reactions

Uracil readily undergoes regular reactions including oxidation, nitration, and alkylation. While in the presence of Phenol/Sodium hypochlorite (PhOH/NaOCl), uracil can be visualized in the blue region of UV light. Uracil also has the capability to react with elemental halogens because of the presence of more than one strongly electron donating group.

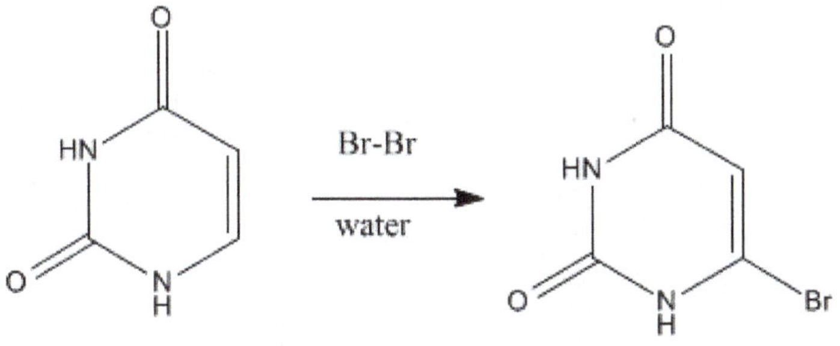

Br-Br

water

Uracil readily undergoes an addition to ribose sugars and phosphates to partake in synthesis and further reactions in the body. Uracil becomes Uridine monophosphate (UMP), uridine diphosphate (UDP), uridine triphosphate (UTP), and uracil diphosphate glucose (UDP-glucose). Each one of these molecules in synthesized in the body and has specific functions.

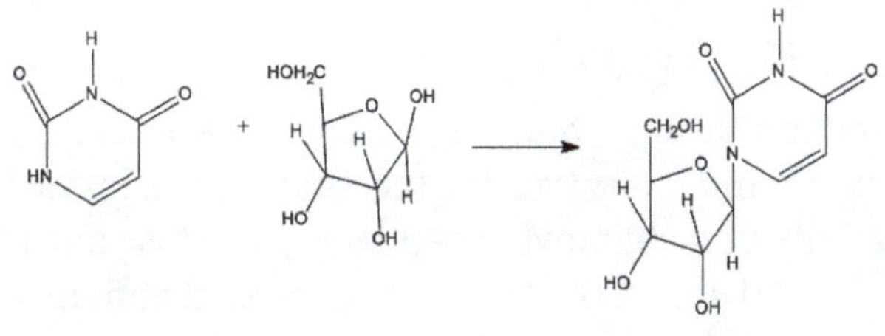

Uridine

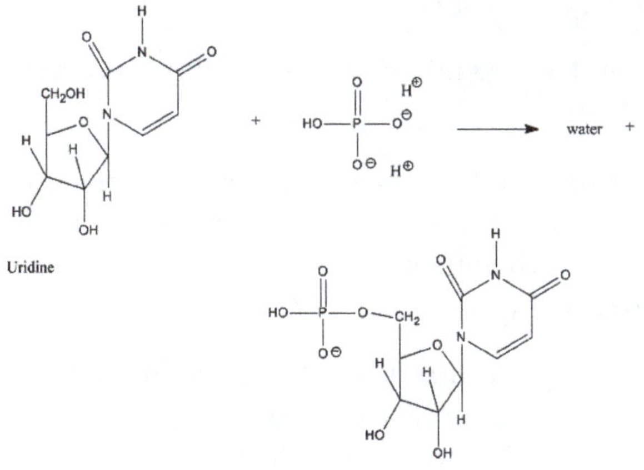

Uridine

water +

Uridine 5' monophosphate

When uracil reacts with anhydrous hydrazine, a first order kinetic reaction occurs and the ring of uracil opens up. If the pH of the reaction increases to >10.5, the uracil anion forms making the reaction go much slower, the same slowing of the reaction occurs if the pH decreases because of the protonation of the hydrazine. The reactivity of uracil is unchanged even if the temperature changes.

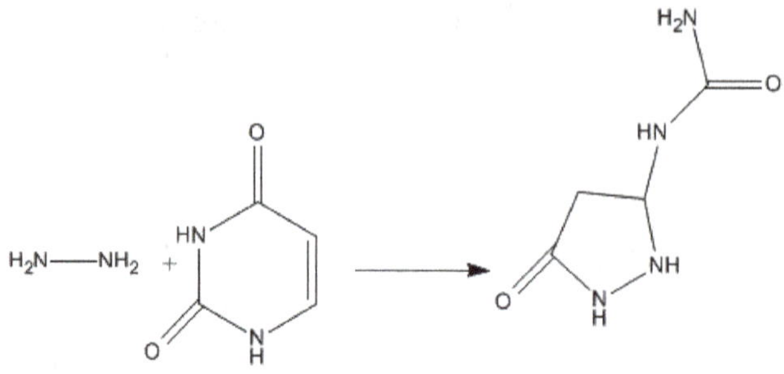

Uses

Uracil can be used for as a pharmaceutical drug with anticancer properties. When elemental fluorine is reacted with uracil, 5-fluorouracil is produced. 5-Fluorouracil is an anticancer drug (antimetabolite) used to masquerade as uracil during the nucleic acid replication process. The molecule also fools the enzymes that help in this process to incorporate this compound in the replication and not uracil, resulting in the biological polymer (cancer) failing to continue synthesizing.

Uracil is used in animals and plants to help carry out the synthesis of many enzymes necessary for cell function through bonding with riboses and phosphates. Uracil serves as an allosteric regulator and coenzyme for reactions in the human body and in plants. UMP controls the activity of carbamoyl phosphate synthetase and aspartate transcarbamoylase in plants, while UDP and UTP requlate CPSase II activity in animals. UDP-glucose regulates the conversion of glucose to galactose in the liverand other tissues in the process of carbohydrate metabolism.

Uracil is also involved in the biosynthesis of polysaccharides and the transportation of sugars containing aldehydes.

Uracil can also increase the risk for cancer in cases where the body is extremely deficient in folate. The deficiency in folate leads to increased ratio of deoxyuracilmonophosphates (dUMP) to deoxythyminemonophosphates (dTMP) and uracil misincorporation into DNA and eventually low production of DNA.

Uracil has a number of agricultural uses. It can be used to determine microbial contamination of tomatoes, since only after lactic acid bacteria have contaminated the fruit does uracil appear. Uracil's derivatives, which contain a diazine ring, are used in pesticides. More often, it is used as an anti-photosynthetic herbicide to destroy weeds among cotton, sugar beet, turnip, soya, pea, and sunflower crops, as well as in vineyards, berry plantations, and orchards.

References

- Mitochondrial-dna-explained-144: phillipsdnaproject.com, Retrieved 22 May 2018
- Mtdna-and-mitochondrial-diseases-903: nature.com, Retrieved 12 July 2018

- What-is-sperm-dna-fragmentation: episona.com, Retrieved 15 June 2018
- Role-of-DNA-Methylation-in-Disease, life-sciences: Retrieved 25 March 2018
- Methods-of-DNA-Shuffling-3385: biotecharticles.com, Retrieved 21 May 2018
- DNA-structure-polynucleotide: byjus.com, Retrieved 18 April 2018

Chapter 4

DNA Sequencing

The process of determining the order of nucleotides within a DNA molecule is known as DNA sequencing. This involves determining the order of adenine, cytosine, guanine and thymine in a DNA molecule. The topics elaborated in this chapter on DNA sequencing theory, next generation sequencing, whole genome sequencing, exome sequencing, polony sequencing, etc. will provide a holistic understanding of DNA sequencing.

DNA sequencing is the process of determining the exact sequence of nucleotides within a DNA molecule. This means that by sequencing a stretch of DNA, it will be possible to know the order in which the four nucleotide bases – adenine, guanine, cytosine and thymine occur within that nucleic acid molecule.

The necessity of DNA sequencing was first made obvious by Francis Crick's theory that the sequence of nucleotides within a DNA molecule directly influenced the amino acid sequences of proteins. At the time, the belief was that a completely sequenced genome would lead to a quantum leap in understanding the biochemistry of cells and organisms.

DNA Sequencing Methods

There are two main types of DNA sequencing. The older, classical chain termination method is also called the Sanger method. Newer methods that can process a large number of DNA molecules quickly are collectively called High-Throughput Sequencing (HTS) techniques or Next-Generation Sequencing (NGS) methods.

Sanger Method

The Sanger method relied on a primer that would bind to a denatured DNA molecule and initiate the synthesis of a single-stranded polynucleotide in the presence of DNA polymerase enzyme, using the denatured DNA as a template. In most circumstances, the enzyme would catalyze the addition of a nucleotide. A covalent bond would therefore form between the 3′ carbon atom of the deoxyribose sugar molecule in one nucleotide and the 5′ carbon atom of the next.

The figure shows, formation of a phosphodiester covalent bond between a guanine and adenine nucleotide, through a condensation reaction. On closer examination, it becomes clear that the reaction occurs between the hydroxyl group on the 3' carbon atom on a deoxyribose sugar and the phosphate moieties in the incoming nucleotide.

A sequencing reaction mixture, however, would have a small proportion of modified nucleotides that cannot form this covalent bond due to the absence of a reactive hydroxyl group, giving rise to the term 'dideoxyribonucleotides', i.e., they do not have a 2' or 3' oxygen atom when compared to the corresponding ribonucleotide. This would terminate the DNA polymerization reaction prematurely. At the end of multiple rounds of such polymerizations, a mixture of molecules of varying lengths would be created.

In the earliest attempts at using the Sanger method, the DNA molecule was first amplified using a labeled primer and then split into four test tubes, each having only one type of ddNTP. That is, each reaction mixture would have only one type of modified nucleotide that could cause chain termination. After the four reactions were completed, the mixture of DNA molecules created by chain termination would undergo electrophoresis on a polyacrylamide gel, and get separated according to their length.

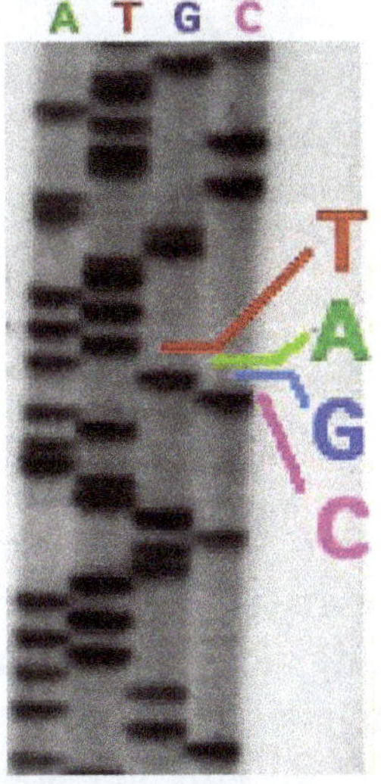

In the figure above, sequencing reaction with ddATP was electrophoresed through the first column. Each line represents a DNA molecule of a particular length, the result of a polymerization reaction that was terminated by the addition of a ddATP nucleotide. The second, third and fourth columns contained ddTTP, ddGTP and ddCTP respectively.

With time, this method was modified so that each ddNTP had a different fluorescent label. The primer was no longer the source of the radiolabel or fluorescent tag. Also known as dye-terminator sequencing, this method used four dyes with non-overlapping emission spectra, one for each ddNTP.

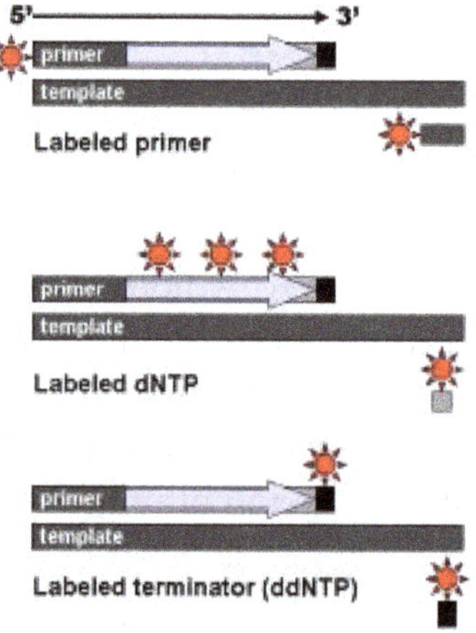

The figure shows difference between labeled primers, labeled dNTPs and dyed terminator NTPs.

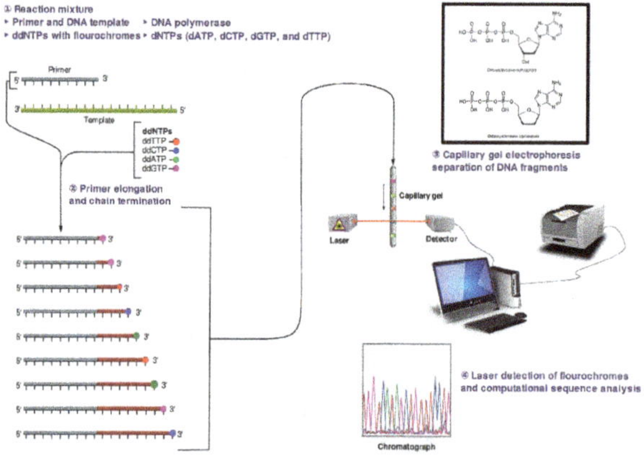

The figure above shows a schematic representation of dye-terminator sequencing. There is a single reaction mixture carrying all the elements needed for DNA elongation. The reaction mixture also contains small concentrations of four ddNTPs, each with a different fluorescent tag. The completed reaction is run on a capillary gel. The results are obtained through an analysis of the emission spectra from each DNA band on the gel. A software program then analyzes the spectra and presents the sequence of the DNA molecule.

High Throughput Sequencing

Sanger sequencing continues to be useful for determining the sequences of relatively long stretches of DNA, especially at low volumes. However, it can become expensive and laborious when a large number of molecules need to be sequenced quickly. Ironically, though the traditional dye-terminator method is useful when the DNA molecule is longer, high-throughput methods have become more widely used, especially when entire genomes need to be sequenced. The human genome project cost nearly 3 billion dollars. In 2004, a large sum of money was pumped into the development of low-cost high-throughput sequencing technology, to potentially allow an entire human genome to be sequenced for less than 1000 dollars.

There were three major changes to the Sanger method. The first was the development of a cell free system for cloning DNA fragments. Traditionally, the stretch of DNA that needed to be sequenced was first cloned into a prokaryotic plasmid, and amplified within bacteria before being extracted and purified. High throughput sequencing or next-generation sequencing technologies no longer relied on this labor-intensive and time-intensive procedure. Secondly, these methods created space to run millions of sequencing reactions in parallel. This was a huge step forward from the initial methods where eight different reaction mixtures were needed to produce a single reliable nucleotide sequence. Finally, there is no separation between the elongation and detection steps. The bases are identified as the sequencing reaction proceeds. While HTS decreased cost and time, their 'reads' were relatively short. That is, in order to assemble an entire genome, intensive computation was necessary, that put together millions of short stretches of sequenced DNA to create the overall nucleotide sequence of a chromosome or genome.

The advent of HTS has vastly expanded the applications for genomics. DNA sequencing has now become an integral part of basic science, translational research, medical diagnostics, and forensics.

Uses of DNA Sequencing

Traditional, chain-termination technology and HTS methods are used for different applications today. Sanger sequencing is now used mostly for *de novo* initial sequencing of a DNA molecule to obtain the primary sequence data for an organism or gene. The relatively short 'reads' coming off a HTS reaction (30-400 base pairs compared to the nearly a thousand base pair 'reads' from Sanger sequencing methods) make it difficult to create the entire genome of any organism from HTS methods alone. Occasionally, Sanger sequencing is also needed to validate the results of HTS.

On the other hand, HTS allows the use of DNA sequencing to understand single-nucleotide polymorphisms – among the most common types of genetic variation within a population. This becomes important in evolutionary biology as well as in the detection of mutated genes that can result in disease. For instance, sequence variations in samples from lung adenocarcinoma allowed the detection of rare mutations associated with the disease. The chromatin binding sites for specific nuclear proteins can also be accurately identified using these methods Overall, DNA sequencing is becoming an integral part of many different applications.

Diagnostics

Genome sequencing is particularly useful for identifying the causes in rare genetic disorders. While more than 7800 diseases are associated with a Mendelian inheritance pattern, less than 4000 of those diseases have been definitively linked to a specific gene or mutation. Early analysis of the exon-genome or exome, consisting of all the expressed genes of an organism, showed promise in identifying the causal alleles for many inherited illnesses. In one particular case, sequencing the genome of a child suffering from a severe form of inflammatory bowel disease connected the illness to a mutation in a gene associated with inflammation – XIAP. While the patient initially showed multiple symptoms suggestive of an immune deficiency, a bone marrow transplant was recommended based on the results of DNA sequencing. The child subsequently recovered from the ailment.

In addition, HTS has been an important player in developing a greater understanding of tumors and cancers. Understanding the genetic basis of a tumor or cancer enables doctors to have an extra tool in their kit for making diagnostic decisions. The Cancer Genome Atlas and International Cancer Genome Consortium have sequenced a large number of tumors and demonstrated that these growths can vary vastly in terms of their mutational landscape. This has also given a better understanding of the kind of treatment options that are ideal for each patient. For instance, the sequencing of the breast cancer genome identified two genes – BRCA1 and BRCA2 – whose pathogenic variants have an enormous impact on the likelihood of developing breast cancer. People with some pathogenic alleles even choose to have preventive surgeries such as double mastectomies.

Molecular Biology

DNA sequencing is now an integral part of most biological laboratories. It is used to verify the results of cloning exercises to understand the effect of particular genes. HTS technologies are used to study variations in the genetic compositions of plasmids, bacteria, yeast, nematodes or even mammals used in laboratory experiments. For instance, a cell line derived from breast cancer tissue, called HeLa is used in many laboratories around the world and was earlier considered as a reliable cell line representing human breast tissue. Recent sequencing results have demonstrated large variations in the genome of HeLa cells from different sources, thereby reducing their utility in cell and molecular biology.

DNA sequencing gives insight into the regulatory elements within the genome of every cell, and the variations in their activity in different cell types and individuals. For instance, a particular gene may be permanently turned off in some tissues, while being constitutively expressed in others. Similarly, those with susceptibility for a specific ailment may regulate a gene differently from those who are immune. These differences in the regulatory regions of DNA can be demonstrated through sequencing and can give insight into the basis for a phenotype.

Recent advances have even allowed individual laboratories to study structural variations in the human genome – an undertaking that needed global collaboration two decades ago.

Forensics

The ability to use low concentrations of DNA to obtain reliable sequencing reads has been extremely useful to the forensic scientist. In particular, the potential to sequence every DNA within a sample is attractive, especially since a crime scene often contains genetic material from multiple people. HTS is slowly being adopted in many forensics labs for human identification. In addition, recent advances allow forensic scientists to sequence the exome of a person after death, especially to determine the cause of death. For instance, death due to poisoning will show changes to the exome in affected organs. On the other hand, DNA sequencing can also determine that the deceased had a preexisting genetic ailment or predisposition. The challenges in this field include the development of extremely reliable analysis software, especially since the results of HTS cannot be manually examined.

Next Generation Sequencing

Next Generation Sequencing (NGS) is a powerful platform that has enabled the sequencing of thousands to millions of DNA molecules simultaneously. This powerful tool is revolutionizing fields such as personalized medicine, genetic diseases, and clinical diagnostics by offering a high throughput option with the capability to sequence multiple individuals at the same time.

The principle behind Next Generation Sequencing (NGS) is similar to that of Sanger sequencing, which relies on capillary electrophoresis. The genomic strand is fragmented, and the bases in each fragment are identified by emitted signals when the fragments are ligated against a template strand.

The Sanger method required separate steps for sequencing, separation (by electrophoresis) and detection, which made it difficult to automate the sample preparation and it was limited in throughput, scalability and resolution. The NGS method uses array-based sequencing which combines the techniques developed in Sanger sequencing to process millions of reactions in parallel, resulting in very high speed and throughput at a reduced cost. The genome sequencing projects that took many years with Sanger methods can now be completed in hours with NGS, although with shorter read lengths (the number of bases that are sequenced at a time) and less accuracy.

Next generation methods of DNA sequencing have three general steps:

- Library preparation: libraries are created using random fragmentation of DNA, followed by ligation with custom linkers.
- Amplification: the library is amplified using clonal amplification methods and PCR.
- Sequencing: DNA is sequenced using one of several different approaches.

Library Preparation

Firstly, DNA is fragmented either enzymatically or by sonication (excitation using ultrasound) to create smaller strands. Adaptors (short, double-stranded pieces of synthetic DNA) are then ligated to these fragments with the help of DNA ligase, an enzyme that joins DNA strands. The adaptors enable the sequence to become bound to a complementary counterpart.

Adaptors are synthesis so that one end is 'sticky' whilst the other is 'blunt' (non-cohesive) with the view to joining the blunt end to the blunt ended DNA. This could lead to the potential problem of base pairing between molecules and therefore dimer formation. To prevent this, the chemical structure of DNA is utilise, since ligation takes place between the 3′-OH and 5′-P ends. By removing the phosphate from the sticky end of the adaptor and therefore creating a 5′-OH end instead, the DNA ligase is unable to form a bridge between the two termini.

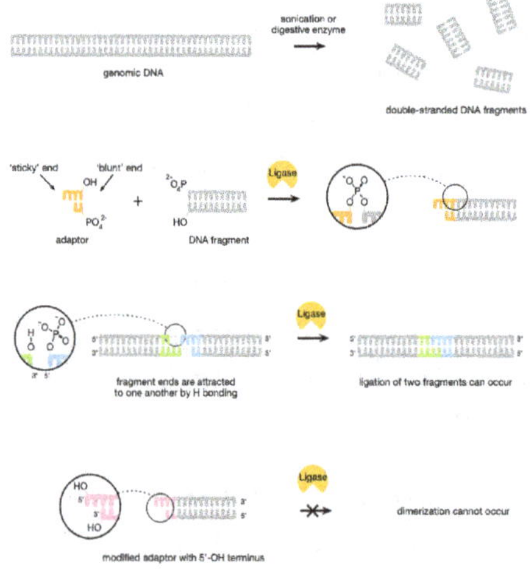

Library preparation of Next-generation sequencing

In order for sequencing to be successful, the library fragments need to be spatially clustered in PCR colonies or 'polonies' as they are conventionally known which consist of many copies of a particular library fragment. Since these polonies are attached in a planar fashion, the features of the array can be manipulated enzymatically in parallel. This method of library construction is much faster than the previous labor intensive procedure of colony picking and E. coli cloning used to isolate and amplify DNA for Sanger sequencing, however, this is at the expense of read length of the fragments.

Amplification

Library amplification is required so that the received signal from the sequencer is strong enough to be detected accurately. With enzymatic amplification, phenomena such as 'biasing' and 'duplication' can occur leading to preferential amplification of certain library fragments. Instead, there are several types of amplification process which use PCR to create large numbers of DNA clusters.

Emulsion PCR

Emulsion oil, beads, PCR mix and the library DNA are mixed to form an emulsion which leads to the formation of micro wells.

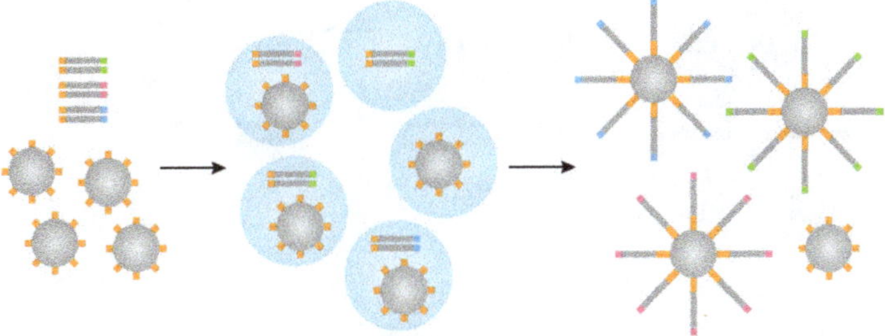

Emulsion PCR

In order for the sequencing process to be successful, each micro well should contain one bead with one strand of DNA (approximately 15% of micro wells are of this composition). The PCR then denatures the library fragment leading two separate strands, one of which (the reverse strand) anneals to the bead. The annealed DNA is amplified by polymerase starting from the bead towards the primer site. The original reverse strand then denatures and is released from the bead only to re-anneal to the bead to give two separate strands. These are both amplified to give two DNA strands attached to the bead. The process is then repeated over 30-60 cycles leading to clusters of DNA. This technique has been criticised for its time consuming nature, since it requires many steps (forming and breaking the emulsion, PCR amplification, enrichment etc.) despite its extensive use in many of the NGS platforms. It is also relatively inefficient since only around two thirds of the emulsion micro reactors will actually contain one bead. Therefore an extra step is required to separate empty systems leading to more potential inaccuracies.

Bridge PCR

The surface of the flow cell is densely coated with primers that are complementary to the primers attached to the DNA library fragments. The DNA is then attached to the surface of the cell at random where it is exposed to reagents for polymerase based extension. On addition of nucleotides and enzymes, the free ends of the single strands of DNA attach themselves to the surface of the cell via complementary primers, creating bridged structures. Enzymes then interact with the bridges to make them double stranded, so that when the denaturation occurs, two single stranded DNA fragments are attached to the surface in close proximity. Repetition of this process leads to clonal clusters of localised identical strands. In order to optimise cluster density, concentrations of reagents must be monitored very closely to avoid overcrowding.

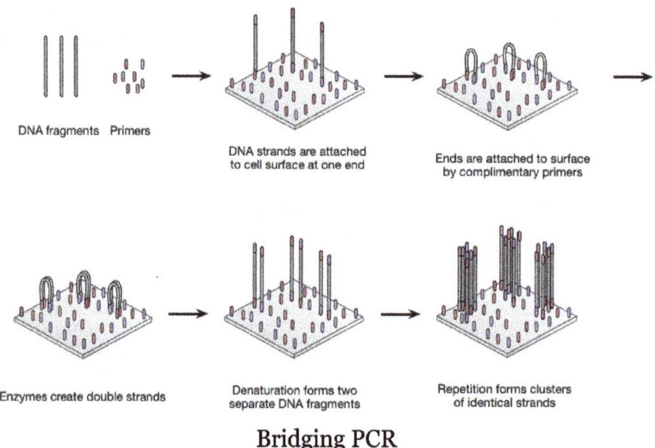

Bridging PCR

Sequencing

Several competing methods of Next Generation Sequencing have been developed by different companies.

Pyro Sequencing

Pyro sequencing is based on the 'sequencing by synthesis' principle, where a complementary strand is synthesis in the presence of polymerase enzyme. In contrast to using dideoxynucleotides to terminate chain amplification (as in Sanger sequencing), pyro sequencing instead detects the release of pyrophosphate when nucleotides are added to the DNA chain. It initially uses the emulsion PCR technique to construct the polonies required for sequencing and removes the complementary strand. Next, ssDNA sequencing primer hybridizes to the end of the strand (primer-binding region), then the four different dNTPs are then sequentially made to flow in and out of the wells over the polonies. When the correct dNTP is enzymatically incorporated into the strand, it causes release of pyrophosphate. In the presence of ATP sulfurylase and adenosine, the pyrophosphate is converted into ATP. This ATP molecule is used for luciferase-catalyse conversion of luciferin to oxyluciferin, which produces light that can be detected with a camera. The relative intensity of light is proportional to the amount of base added (i.e. a peak of twice the intensity indicates two identical bases have been added in succession).

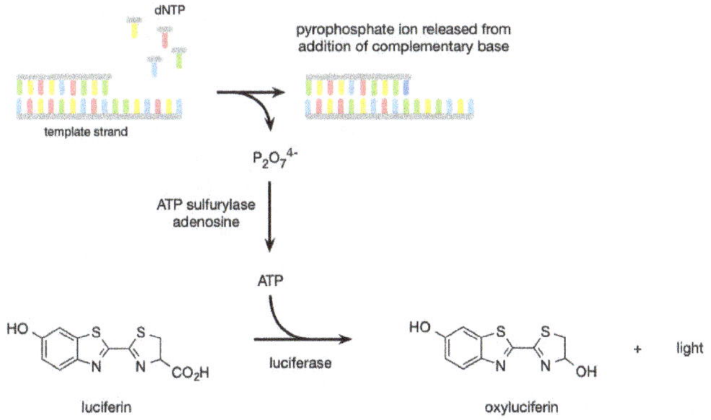

Pyro sequencing

Pyro sequencing, developed by Life Sciences, was one of the early successes of Next-generation sequencing; indeed, Life Sciences produced the first commercially available Next-generation sequencer. However, the method was eclipsed by other technologies and in 2013 new owners Roche announced the closure of Life Sciences and the discontinuation of the pyro sequencing platform.

Ion Torrent Semiconductor Sequencing

Ion torrent sequencing uses a "sequencing by synthesis" approach, in which a new DNA strand, complementary to the target strand, is synthesized one base at a time. A semiconductor chip detects the hydrogen ions produced during DNA polymerization.

Following polony formation using emulsion PCR, the DNA library fragment is flooded sequentially with each nucleoside triphosphate (dNTP), as in pyro sequencing. The dNTP is then incorporated into the new strand if complementary to the nucleotide on the target strand. Each time a nucleotide is successfully added, a hydrogen ion is released, and it detected by the sequencer's pH sensor. As in the pyro sequencing method, if more than one of the same nucleotide is added, the change in pH/signal intensity is correspondingly larger.

Ion torrent sequencing is the first commercial technique not to use fluorescence and camera scanning; it is therefore faster and cheaper than many of the other methods. Unfortunately, it can be difficult to enumerate the number of identical bases added consecutively. For example, it may be difficult to differentiate the pH change for a homo repeat of length 9 to one of length 10, making it difficult to decode repetitive sequences.

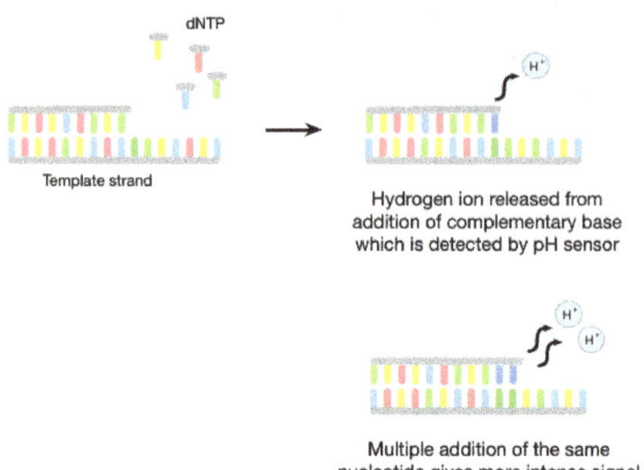

Ion Torrent semiconductor sequencing

Sequencing by Ligation (SOLiD)

SOLiD is an enzymatic method of sequencing that uses DNA ligase, an enzyme used widely in biotechnology for its ability to ligate double-stranded DNA strands. Emulsion PCR is used to immobilize / amplify ssDNA primer-binding region (known as an adapter) which has been conjugated to the target sequence (i.e. the sequence that is to be sequenced) on a bead. These beads are then deposited onto a glass surface – a high density of beads can be achieved which in turn, increases the throughput of the technique.

Once bead deposition has occurred, a primer of length N is hybridized to the adapter then the beads are exposed to a library of 8-mer probes which have different fluorescent dye at the 5' end and a hydroxyl group at the 3' end. Bases 1 and 2 are complementary to the nucleotides to be sequenced whilst bases 3-5 are degenerate and bases 6-8 are in osine bases. Only a complementary probe will hybridize to the target sequence, adjacent to the primer. DNA ligase is then uses to join the 8-mer probe to the primer. A phosphorothioate linkage between bases 5 and 6 allows the fluorescent dye to be cleaved from the fragment using silver ions. This cleavage allows fluorescence to be measured (four different fluorescent dyes are used, all of which have different emission spectra) and also generates a 5'-phosphate group which can undergo further ligation. Once the first round of sequencing is completed, the extension product is melted off and then a second round of sequencing is perfomed with a primer of length N−1. Many rounds of sequencing using shorter primers each time (i.e. N−2, N−3 etc.) and measuring the fluorescence ensures that the target is sequenced.

Due to the two-base sequencing method (since each base is effectively sequenced twice), the SOLiD technique is highly accurate (at 99.999% with a sixth primer, it is the most accurate of the second generation platforms) and also inexpensive. It can complete a single run in 7 days and in that time can produce 30 Gb of data. Unfortunately, its main disadvantage is that read lengths are short, making it unsuitable for many applications.

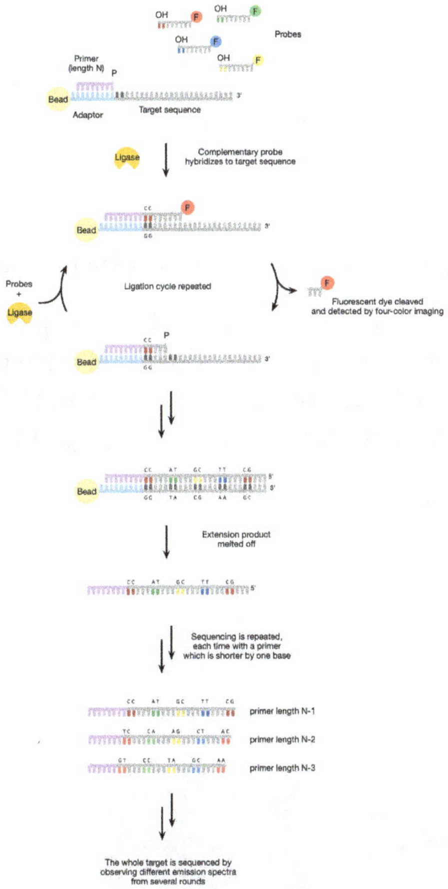

Sequencing by ligation

Reversible Terminator Sequencing (Illumina)

Reversible terminator sequencing differs from the traditional Sanger method in that, instead of terminating the primer extension irreversibly using dideoxynucleotide, modified nucleotides are used in reversible termination. Whilst many other techniques use emulsion PCR to amplify the DNA library fragments, reversible termination uses bridge PCR, improving the efficiency of this stage of the process.

Reversible terminators can be grouped into two categories: 3′-O-blocked reversible terminators and 3′-unblocked reversible terminators.

3′-O-Blocked Reversible Terminators

The mechanism uses sequencing by synthesis approach, elongating the primer in a stepwise manner. Firstly, the sequencing primers and templates are fixed to a solid support. The support is exposed to each of the four DNA bases, which have a different fluorophore attached (to the nitrogenous base) in addition to a 3’-O-azidomethyl group.

Structure of fluorescently labelled dNTP used in Illumina sequencing

Only the correct base anneals to the target and is subsequently ligated to the primer. The solid support is then imaged and nucleotides that have not been incorporated are washed away and the fluorescent branch is cleaved using TCEP (tris(2-carboxyethyl) phosphine). TCEP also removes the 3'-O-azidomethyl group, regenerating 3'-OH, and the cycle can be repeated.

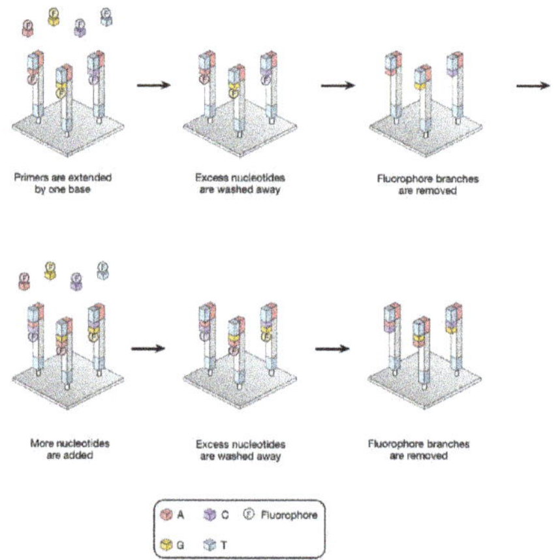

Reversible terminator sequencing

3′-Unblocked Reversible Terminators

The reversible termination group of 3′-unblocked reversible terminators is linked to both the base and the fluorescence group, which now acts as part of the termination group as well as a reporter. This method differs from the 3′-O-blocked reversible terminators method in three ways: firstly, the 3'-position is not blocked (i.e. the base has free 3'-OH); the fluorophore is the same for all four bases; and each modified base is flowed in sequentially rather than at the same time.

The main disadvantage of these techniques lies with their poor read length, which can be caused by one of two phenomena. In order to prevent incorporation of two nucleotides in a single step, a block is put in place, however in the event of no block addition due to a poor synthesis, strands can become out of phase creating noise which limits read length. Noise can also be created if the fluorophore is unsuccessfully attached or removed. These problems are prevalent in other sequencing methods and are the main limiting factors to read length.

This technique was pioneered by Illumina, with their HiSeq and MiSeq platforms. HiSeq is the cheapest of the second generation sequencers with a cost of $0.02 per million bases. It also has a high data output of 600 GB per run which takes around 8 days to complete.

Third Generation Sequencing

A new cohort of techniques has since been developed using single molecule sequencing and single real time sequencing, removing the need for clonal amplification. This reduces errors caused by PCR, simplifies library preparation and, most importantly, gives a much higher read length using higher throughput platforms. Examples include Pacific Biosciences' platform which uses SMRT (single molecule real time) sequencing to give read lengths of around one thousand bases and Helicos Biosciences which utilises single molecule sequencing and therefore does not require amplification prior to sequencing. Oxford Nanopore Technologies are currently developing silicon-based nanopores which are subjected to a current that changes as DNA passes through the pore. This is anticipated to be a high-throughput rapid method of DNA sequencing, although problems such as slowing transportation through the pore must first be addressed.

Sequencing Epigenetic Modifications

Just as Next generation sequencing enabled genomic sequencing on a massive scale it has become clear recently that the genetic code does not contain all the information needed by organisms. Epigenetic modifications to DNA bases, in particular 5-methylcytosine, also convey important information.

All of the second generation sequencing platforms depend, like Sanger sequencing, on PCR and therefore cannot sequence modified DNA bases. In fact, both 5-methylcytosine and 5-hydroxymethylcytosine are treated as cytosine by the enzymes involved in PCR; therefore, epigenetic information is lost during sequencing.

Bisulfite Sequencing

Bisulfite sequencing exploits the difference in reactivity of cytosine and 5-methylcytosine with respect to bisulfite: cytosine is deaminated by bisulfite to form uracil (which reads as T when sequenced), whereas 5-methylcytosine is unreactive (i.e. reads as C). If two sequencing runs are done in parallel, one with bisulfite treatment and one without, the differences between the outputs of the two runs indicate methylated cytosines in the original sequence. This technique can also be used for dsDNA, since after treatment with bisulfite the strands are no longer complementary and can be treated as ssDNA.

5-Hydroxymethylcytosine, another important epigenetic modification, reacts with bisulfite to form cytosine-5-methylsulfonate (which reads as C when sequenced). This complicates matters somewhat, and means that bisulfite sequencing cannot be used as a true indicator of methylation in itself.

Oxidative Bisulfite Sequencing

Oxidative bisulfite sequencing adds a chemical oxidation step, which converts 5-hydroxymethylcytosine to 5-formylcytosine using potassium perruthenate, KRuO4, before bisulfite treatment. 5-Formylcytosine is deformylated and deaminated to form uracil by bisulfite treatment. Now, three separate sequencing runs are necessary to distinguish cytosine, 5-methylcytosine and 5-hydroxymethylcytosine.

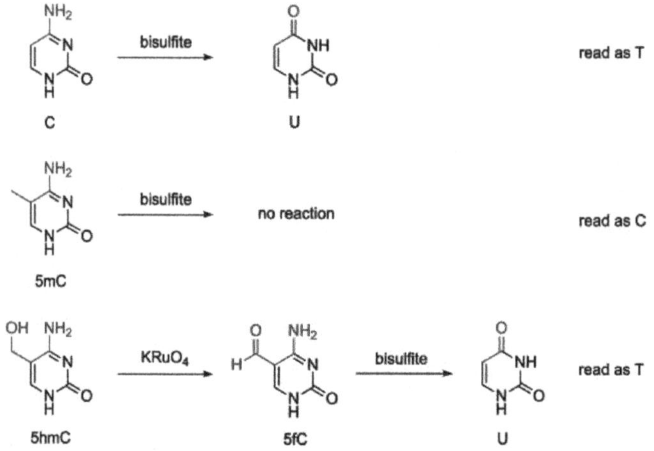

Sequencing epigenetic modifications using bisulfite

Applications of Next-Generation Sequencing

Next generation sequencing has enabled researchers to collect vast quantities of genomic sequencing data. This technology has a plethora of applications, such as: diagnosing and understanding complex diseases; whole-genome sequencing; analysis of epigenetic modifications; mitochondrial sequencing, transcriptome sequencing – understanding how altered expression of genetic variants affects an organism; and exome sequencing – mutations in the exome are thought to contain up to 90% of mutations in the human genome, which leads to disease. DNA techniques have been used to identify and isolate genes responsible for certain diseases, and provide the correct copy of the defective gene known as 'gene therapy'.

A large focus area in gene therapy is cancer treatment – one potential method would be to introduce an antisense RNA (which specifically prevents the synthesis of a targeted protein) to the oncogene, which is triggered to form tumorous cells. Another method is named 'suicide gene therapy' which introduces genes to kill cancer cells selectively. Many genetic codes for toxic proteins and enzymes are known, and introduction of these genes into tumor cells would result in cell death. The difficulty in this method is to ensure a very precise delivery system to prevent killing healthy cells.

These methods are made possible by sequencing to analyze tumor genomes, allowing medical experts to tailor chemotherapy and other cancer treatments more effectively to their patients' unique genetic composition, revolutionizing the diagnostic stages of personalized medicine.

As the cost of DNA sequencing goes down, it will become more widespread, which brings a number of issues. Sequencing produces huge volumes of data, and there are many computational challenges associated with processing and storing the data. There are also ethical issues, such as the ownership of an individual's DNA when the DNA is sequenced. DNA sequencing data must be stored securely, since there are concerns that insurance groups, mortgage brokers and employers may use this data to modify insurance quotes or distinguish between candidates. Sequencing may also help to find out whether an individual has an increased risk to a particular disease, but whether the patient is informed or if there is a cure for the disease is another issue altogether.

Whole Genome Sequencing

Whole genome sequencing (WGS) is performed on human samples, model organisms, and microbial organisms. Deep coverage whole genome sequencing of human and model organisms is the ultimate tool for discovering the full range of genetic variations, including SNPs, insertions, deletions, inversions, complex rearrangements, and copy number variations. Low-pass whole genome human sequencing can also be performed to assess genomic variations such as aneuploidy, where large regions or whole chromosomes can be duplicated or deleted.

Whole genome sequencing of microbial organisms is used to characterize and discover new organisms or to type specific bacterial organisms.

Generating deep coverage of whole genomes for complex organisms requires technologies capable of sequencing in a high-throughput and cost-effective manner. Applications like aneuploidy detection require less depth in coverage (low-pass coverage) of the entire genome and benefit from clear and precise data generated in a single-day workflow at an affordable price.

A Complete View of Genetic Variation

Analyzing the whole genome using next-generation sequencing (NGS) delivers a base-by-base view of all genomic alterations, including single nucleotide variants (SNV), insertions and deletions, copy number changes, and structural variations. Paired-end whole-genome sequencing involves sequencing both ends of a DNA fragment, which increases the likelihood of alignment to the reference and facilitates detection of genomic rearrangements, repetitive sequences, and gene fusions.

Advantages of Large Whole-Genome Sequencing

- Provides a high-resolution, base-by-base view of the genome;
- Combines short inserts and longer reads to allow characterization of any genome;
- Reveals disease-causing alleles that might not have been identified;
- Identifies potential causative variants for further follow-on studies of gene expression and regulation mechanisms.

Genotyping by Sequencing

Advances in NGS have driven the costs of DNA sequencing down to the point that GBS is now feasible for high diversity and large genome species. GBS is a simple highly multiplexed system for constructing reduced representation libraries for the Illumina NGS platform developed in the Buckler lab. It generates large numbers of SNPs for use in genetic analyses and genotyping. Key components of this system include low cost, reduced sample handling, fewer PCR and purification steps, no size fractionation, no reference sequence limits, efficient barcoding and easiness to scale up. GBS is becoming increasingly important as a cost-effective and unique tool for genomics-assisted breeding in a range of plant species. Figure simplifies the GBS technology by summarizing the steps needed for any plant species and some potential application of the results.

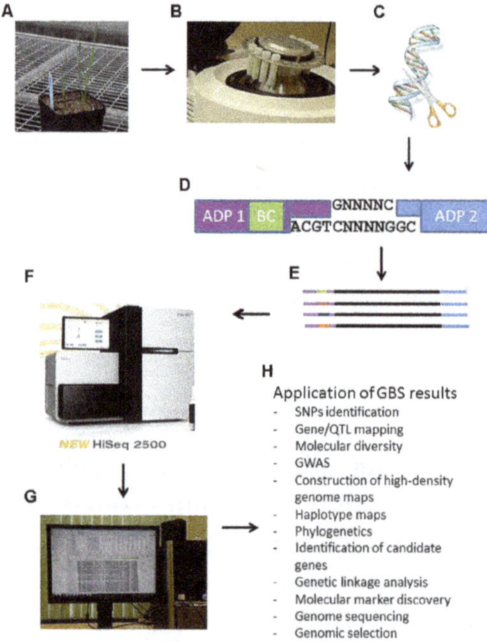

Schematic steps of the genotyping-by-sequencing (GBS) protocol for plant breeding.

Panel (A): tissue is obtained from any plant species as depicted here a young triticale plant;

Panel (B): ground leaf tissues for DNA isolation, quantification and normalization. At this step it is important to prevent any cross-contamination among samples;

Panel (C): DNA digestion with restriction enzymes;

Panel (D): ligations of adaptors (ADP) including a bar coding (BC) region in adapter 1 in random *PstI-Mse*I restricted DNA fragments;

Panel (E): representation of different amplified DNA fragments with different bar codes from different biological samples/lines. These fragments represent the GSB library;

Panel (F): analysis of sequences from library on a NGS sequencer;

Panel (G): bioinformatics analysis of NGS sequencing data;

Panel (H): possible application of GBS results.

GBS combined with genome-independent imputation provides a simple and efficient method for genetic map construction in any pseudo-testcross progeny. The GBS method offers a greatly simplified library production procedure more amenable to use on large numbers of individuals/lines. A two-enzyme (*PstI/Msp*I) GBS protocol, which provides a greater degree of complexity reduction and uniform library for sequencing than the original protocol using *Ape*KI, has now been developed and applied to both wheat and barley, describe a modified library preparation protocol, in which selective amplification is used to increase both the number of SNPs called and their depth of coverage, resulting in a high efficiency to allow an important reduction in per sample cost.

Two different GBS strategies have been developed with the Ion PGM system:

A. Restriction enzyme digestion, in which no specific SNPs have been identified and ideal for discovering new markers for MAS programs. The complexity of the genome under this approach is reduced by digesting the DNA with one or two selected restriction enzymes prior to the ligation of the adapters.

B. Multiplex enrichment PCR, in which a set of SNPs has been defined for a section of the genome. This approach uses PCR primers designed to amplify the areas of interest.

First, the value of sequencing restriction site associated genomic DNA (RAD) for high density SNP discovery and genotyping. Increased efficiency and cost benefits were realized by incorporating a multiplex sequencing strategy that uses an inexpensive barcoding system. Barcodes are included in one of the adapter sequences, and their locations, just upstream of the RE cut-site in genomic DNA, eliminate the need for a second Illumina sequencing ("indexing") read. The barcoding strategy is similar to RAD but modulation of barcode nucleotide composition and length results in fewer sequence phasing errors. Compared to the RAD method, GBS is substantially less complicated; generation of restriction fragments with appropriate adapters is more straightforward, single-well digestion of genomic DNA and adapter ligation results in reduced sample handling, there are fewer DNA purification steps, and fragments are not size selected. Costs can be further reduced via shallow genome sampling coupled with imputation of missing internal SNPs in haplotype blocks.

GBS was originally developed for high resolution association studies in maize and, like RAD, has been extended to a range of species with complex genomes. Unlike other high density genotyping technologies which have mainly been applied to general interest "reference" genomes, the low cost of GBS makes it an powerful approach on discovering and genotyping SNPs in a variety of crop species and populations. As a technically simple, highly multiplexed technology, GBS is suitable for population studies, germplasm characterization, plant genetics, and breeding in diverse crops and it has widely been applied in many large crop genomes to saturate the mapping and breeding populations with 10–100s of 1000s of SNP markers.

Construction of GBS libraries is based on reducing genome complexity with restriction enzymes. This approach is simple, quick, extremely specific, highly reproducible, and may reach important regions of the genome that are inaccessible to sequence capture approaches. By choosing appropriate REs, repetitive regions of genomes can be avoided, and lower copy regions can be targeted with two to three fold higher efficiency, which tremendously simplifies computationally challenging alignment problems in species with high levels of genetic diversity. The GBS procedure is demonstrated with maize and barley recombinant inbred populations where roughly 200,000 and 25,000 sequence tags were mapped, respectively.

Application of GBS in Plant Breeding

Genotyping-by-sequencing is an ideal platform for studies ranging from single gene markers to whole genome profiling. One of the most powerful applications of GBS is in the field of plant breeding. GBS provides a rapid and low-cost tool to genotype breeding populations, allowing plant breeders to implement GWAS, genomic diversity study, genetic linkage analysis, molecular marker discovery, and genomic selection (GS) under a large scale of plant breeding programs. There is no requirement for a priori knowledge of the species genomes as the GBS method has been shown to be robust across a range of species and SNP discovery and genotyping are completed together.

As GWAS require 100s of 1000s to millions of markers to generate sufficient information and coverage, the emergence of NGS technologies has greatly improved such marker resolution. Recently, GBS through the NGS approach has been used to re-sequence collections of recombinant inbred lines (RILs) to analyze and map various traits of interest in specific breeding programs. More and more crops, such as maize, wheat, barley, rice, potato, and cassava, have been optimized by GBS for the efficient, low-cost and large scales of genome sequencing. A collection of 5,000 RILs have been re-sequenced using a restriction endonuclease-based approach and the Illumina sequencing technology, which generated a total of 1.4 million SNPs and 200,000 indels in maize. A comprehensive genotyping of 2,815 maize inbred accessions showed that 681,257 SNP markers are distributed across the entire genome, in which some SNPs are linked to the known candidate genes for kernel color, sweetness, and flowering time. A set of 205,614 SNPs have been identified after re-sequencing 31 soybean genotypes, providing a valuable genomic resource for soybean breeding programs. In potato, 12.4 gigabases of high-quality sequence data and 129,156 sequence variants have been identified, which are mapped to 2.1 Mb of the potato reference genome with a median average read depth of 636 per cultivar.

GBS has been shown to be a valid tool for genomic diversity studies. For example, Fu and Peterson (2011) applied the Roche 454 GS FLX Titanium technology with reduced genome representation and advanced bioinformatics tools to analyze the genetic diversity of 16 diverse barley landraces, discovered 2,578 contigs and 3,980 SNPs, and confirmed a key geographical division in the cultivated barley gene pool, developed a network-based SNP discovery protocol to enhance the diversity analysis of 540 switch grass plants sampled from 66 populations and revealed informative patterns of genetic relationship with respect to ecotype, ploidy level, and geographic distribution. The GBS protocol was used to analyze genetic diversity of 24 diverse yellow mustard accessions, in which roughly 1.2 million sequence reads (total about 392 million nucleotides) were generated, 512 contigs, and 828 SNPs were identified. Diversity analysis of these yellow mustard SNPs revealed that 26.1% of total variation resided among landrace, cultivar, and breeding lines and 24.7% between yellow-seeded and black-seeded germplasm.

Identification of high density SNP markers through GBS to construct genetic linage maps has a great value for numerous applications in plant breeding. In *Arabidopsis*, sequenced, via whole genome shotgun sequencing on the Illumina platform, a pool of 500 F2 plants generated by crossing a recessive ethane methyl sulfonate (EMS)-induced Col-0 mutant characterized by slow growth and light green leaves, with a wild type Ler(Landsberg *erecta*) line used a 384 plex GBS protocol to add 30,984 SNP markers to an *indica* × *japonica* mapping population consisting of 176 rice recombinant inbred lines and mapped the recombined hot and cold spots and quantitative trait loci (QTLs) for leaf width and aluminum tolerance. After the efficiency of multiplexed SNP genotyping for diversity, mapping and breeding applications were evaluated, Thomson et al. demonstrated that 384 plex SNP genotyping on the Bead Xpress platform is a robust and efficient method for marker genotyping and mapping in rice. GBS was applied to bread wheat, resulting in the incorporation of 1000s of markers in the bread wheat map. The high resolution of SNP markers were identified in barley and the GBS mapping data were used to confirm that the semi-dwarfing gene is located on barley chromosome 5H. Construction of a GBS linkage map using the sequence-based markers leads to the RAD technique, which has been used in barley QTL analysis.

By integrating molecular markers and genotyping of large populations, GBS is an excellent platform for plant breeding applications even in the absence of reference genome sequences or without previous DNA polymorphism discovery. The GBS approach has been shown to be suited to genetic analysis and marker development of rapeseed, lupin, lettuce, switch grass, soybean, and maize. With Illumina genome analyzer, Varala et al. identified 4294 to 14550 SNPs between four soybean accessions and the reference and indicated that the *Mse*I digestion of soybean genomic DNA followed by high throughput sequencing provides a rapid and reproducible method for generating SNP markers. High-throughput SNP discovery and genotyping in durum wheat have been investigated from 92 RILs derived from a cross between the two elite cultivars. The application of GBS on a large collection of autotetraploid potato cultivars were studied with Illumina HiSeq2000 and the alleles strongly associating with maturity and flesh color were identified.

Compared to traditional MAS, GS is a novel approach which combines molecular markers with phenotypic and pedigree data to increase the breeding accuracy on genotypic values. Theoretical and applied studies on GS show great promise to accelerate the rate of developing new crop varieties. GS through the GBS approach stands to be a major supplement to traditional crop improvement and it is a very important feature to move the genomics-assisted breeding into commercial crops with large and complex genomes. One premise of GBS applications is the development of genome-wide molecular markers with high density and low cost. GBS approach on barley and wheat study provides a powerful method of developing high density markers in species without a reference genome while providing valuable tools for anchoring and ordering physical maps and whole genome shotgun sequence used GBS to discover 41,371 SNPs in a set of 254 advanced breeding lines from CIMMYT's semiarid wheat breeding program reported that the high marker density allows the identification of genomic regions with segregation distortion in *Rubus idaeus*, which may help to identify deleterious alleles that are the basis of inbreeding depression in that species. An efficient GBS approach has been developed to catalog SNPs both within the mapping population and among diverse African cassava varieties, allowing the improvements of MAS programs on disease resistance and nutrition in cassava.

Although GBS offers a novel approach on enhancing the efficiency and capacity of plant breeding, some potential drawbacks have been identified under its applications, which seems not unique to this technique. A major challenge encountered by all genotyping methods has been the difficulty to align true alleles of each single locus in large, complex, polyploidy genomes. Among all the tools available, however, GBS is the one offering the higher potential to resolve the issue. As exemplified by alleles in hexaploid oat can be distinguished after extensive analyses of sequence data through two different bioinformatics pipelines, suggesting the data analysis algorithms may now represent the limiting factor to ascertain alleles at each single locus in a large polyploidy genome rather than GBS itself given sufficient depth of sequence is available.

The reduction in genome complexity using restriction enzymes in the GBS protocol means that, in case of any mutation at the restriction site, the genomic DNA of this region is not available to be PCR amplified and consequently the SNPs of this region will become unavailable. In the worst case of this scenario, a heterozygote gene may appear as homozygous. However, this is a drawback shared by all the different methods involving reduction in genome complexity based on the utilization of restriction sites. No scientist is ready to sacrifice the high throughput of these methods to move back to RFLP-based protocols. The feasibility of reduced representation and highly multiplexed GBS strategy was demonstrated in the complex genomes of maize and barley via a simple procedure targeting regions flanking restriction endonuclease sites.

Epigenetic studies have revealed the importance of differential DNA methylation in numerous biological systems. Two restriction enzymes (CTGCAG), were employed to improve the reduction of genome complexity in barley and wheat. It may have been overlooked that the activity of *Msp*I is inhibited when the DNA is methylated at the external "C." In epigenetic studies, however, the activity of the isoschizomer *Hpa*II is inhibited by methylation at any of two "C." Because of the wide utilization of *Hpa*II, the methylation of the internal "C" in epigenetic regulation studies seems to be much more important than the methylation of the external "C." Therefore, the possibility that developmental responses in plants may affect the SNP identification when using the enzyme *Msp*I cannot be ignored, but is likely reduced.

Orphan plant species without a known genomic sequence represent the vast majority of crops over the world. The GBS protocol for wheat and barley and subsequent genetic analyses were carried out when a draft genomic sequence was not available yet. An available reference genome can simplify the data analyses, but it is not essential in GBS, indicating a great advantage of the GBS technique in accelerating plant breeding and crop improvement. This reality has been confirmed with the recent GBS applications on different oat accessions. The depth of genomic sequencing is important to identify stable and representative SNPs which can be generated to improve crop genotypes also demonstrate the importance of the bioinformatics pipeline to fully exploit the GBS datasets, which is likely more critical in orphan plant species.

Advantages

- No prior knowledge of the genome sequence required:
- Discover new SNPs in your species without the need for a reference genome.
- Cost effective Find thousands of markers in any sample number:

- From a handful of samples for a few million reads to thousands of samples multiplexed on high-throughput NGS platforms.
- Professional service and rapid turnaround times:
- Licensed technology for all species, unique service provider for vegetable crops.
- Compatible with high-throughput KASP genotyping:
- Easy conversion of identified markers to LGC's flexible and high-throughput KASP assays.
- Bioinformatics support for any species:
- Including sequence pre-processing, alignment and variant analysis.

Exome Sequencing

Perhaps the most widely used targeted sequencing method is exome sequencing. The exome (the protein-coding region of the human genome) represents less than 2% of the genome, but contains ~85% of known disease-related variants, making whole exome sequencing a cost-effective alternative to whole-genome sequencing.

Exome sequencing can efficiently identify coding variants across a wide range of applications, including population genetics, genetic disease, and cancer studies.

Efficient Analysis of Coding Regions

Exome sequencing is a cost-effective approach when whole-genome sequencing is not practical or necessary. Sequencing only the coding regions of the genome enables researchers to focus their resources on the genes most likely to affect phenotype, and offers an accessible combination of turnaround time and price.

Exome sequencing detects variants in coding exons, with the capability to expand targeted content to include untranslated regions (UTRs) and microRNA for a more comprehensive view of gene regulation. DNA libraries can be prepared in as little as 1 day and require only 4–5 GB of sequencing per exome.

Advantages of Exome Sequencing

- Identifies variants across a wide range of applications;
- Achieves comprehensive coverage of coding regions;
- Provides a cost-effective alternative to whole-genome sequencing (4–5 Gb of sequencing per exome compared to ~90 Gb per whole human genome);

- Produces a smaller, more manageable data set for faster, easier analysis compared to whole-genome approaches.

Solid Phase Sequencing

The solid-phase sequencing principle can best be illustrated by determination of the primary structure of DNA fragments. Used for immobilization is what man 540 paper activated with cyanuric chloride and then converted with 2-bromoethylamine in the presence of triethy lamine.

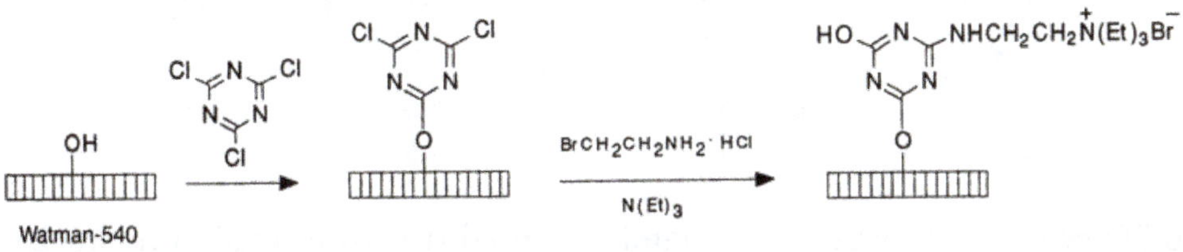

Watman-540

$BrCH_2CH_2NH_2 \cdot HCl$

$N(Et)_3$

The immobilization of DNA on paper prepared in the above manner is achieved through ion exchange. 120 mg of DNA can be immobilized per square centimeter of paper. Depending on the number of DNA samples to be sequenced, 4n pieces of paper (2-4 mm², shown in left figure) are cut and marked with a pencil. Paper strips can also be used (shown in right figure). The DNA to be analyzed (0.5-1.0 ml of sample labeled at the 5' or 3' end) is applied onto the prepared paper. When paper strips are used, the DNA application points must be spaced 2 to 5 mm apart. The paper (all pieces or strips) is placed in four reaction vessels to conduct the reactions (G, G+A, T, C).

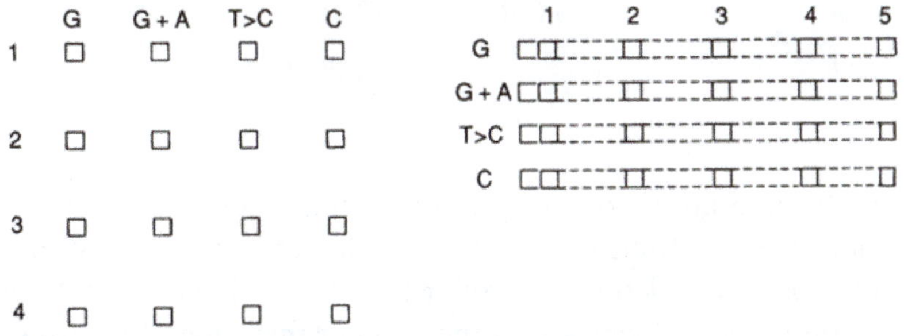

G G+A T>C C

1 2 3 4

1 2 3 4 5

G G+A T>C C

Preparation of ion-exchange paper for immobilization of 5 DNA samples.

The base modification conditions have been changed slightly in comparison with the standard reactions normally used for sequencing in solution. Direct use of the latter for solid-phase sequencing has turned out to be impossible because of the sizable DNA losses during the modifications (hydrazinolysis, alkaline treatment). Dimethyl sulfate is used to modify DNA at the guanines; the position of adenine and guanines (A+G) is determined by treating the samples with 66-88% formic acid or diethyl pyro carbonate at 90° C (A>G). The following scheme illustrates the transformations undergone by adenine in the presence of diethyl pyro carbonate with subsequent opening of the imidazole ring. The first step is acylation of the nitrogen (N7) in imidazole.

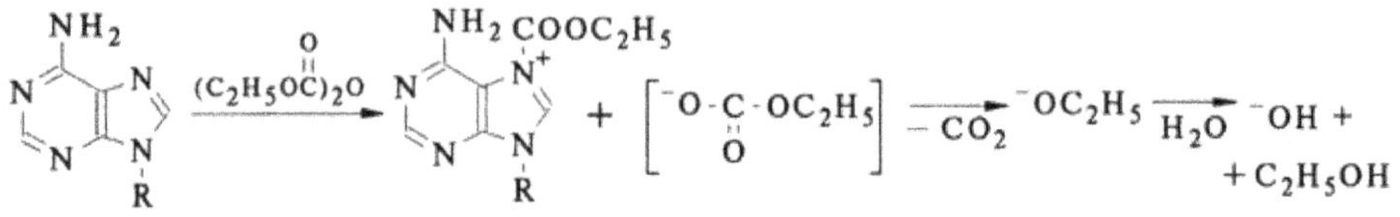

$$\xrightarrow{(C_2H_5OC(O))_2O} \; + \left[{}^-O\text{-}C(O)\text{-}OC_2H_5\right] \xrightarrow{-CO_2} {}^-OC_2H_5 \xrightarrow{H_2O} {}^-OH + C_2H_5OH$$

Subsequently, the hydroxyl anion (strong nucleophile) immediately attacks the carbon at position 8 of the imidazole ring, which leads to cleavage of the C8-C9 bond:

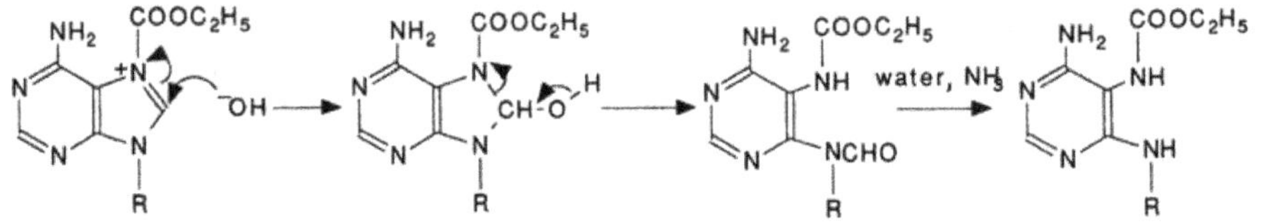

To modify cytosine in DNA hydroxylamine is used (pH 6) instead of hydrazine. It is believed that hydroxylamine is added at the double C5-C6 bond and the amine group is substituted. The base modified in this manner is detached in the presence of piperidine from the polynucleotide chain which is then cleaved as usual.

To determine the position of thymines (T) during solid-phase sequencing use is made of oxidation with potassium permanganate in aqueous solutions. The intermediate products of this reaction are 5,6-dihydroxy-5,6-dihydro derivatives.

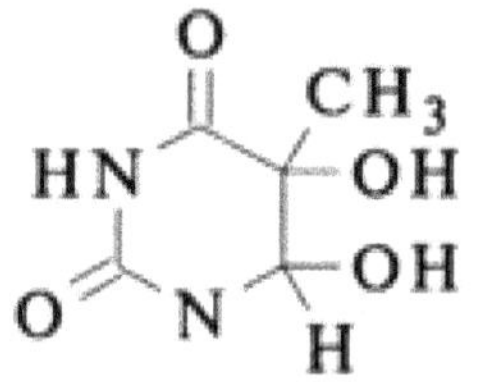

These compounds undergo fast opening of the pyrimidine ring at the C5-C6 bond, and the sequence of transformations results in formation of N-substituted ureas NH2-CO-NH-R (where R is a polynucleotide). Similar transformations are undergone by cytosines. The oxidation rate of thymines is much higher than that of cytosines. Therefore, the reaction with permanganate makes it possible to determine the position of the former in the first place.

Treatment with piperidine (conditions under which β-elimination takes place) is performed separately for each DNA sample (if paper strips are used, they must be cut). The paper is treated with 10 % aqueous piperidine (30 min, 90° C), which leads to elution of the DNA. The subsequent treatment, PAGE and autoradiography are carried out under standard conditions.

It should be noted that the solid-phase method allows one to analyze a large number of DNA samples at a time, which cuts down the sequencing time considerably.

For solid-phase RNA sequencing ion-exchange what man DE 81 paper is used. 3'-Labeled RNA is immobilized on 4n 2-4 MM2 pieces of paper, depending on the number of samples, which are then marked with a pencil and placed in four vessels to perform the corresponding modifications. The following conditions are maintained to determine the position of bases in the RNA chain:

G: dimethyl sulfate (pH 5.5, 90°, 1 min) with subsequent NaBH4 treatment (pH 7, 0°, 30 min)

A>G: diethyl pyrocarbonate (pH 4.5, 90°, 10 min)

U+C: 50 % aqueous hydrazine (0°, 20 min)

C: 3 M NACI in anhydrous hydrazine (0°, 10 min).

After modification all paper segments are treated together with 1 M aniline-acetate buffer (pH 4.5, 60° C, 20 min). For desorption of RNA each paper segment is treated with 2 M $NaCIO_4$. RNA is precipitated with an alcohol and subjected to PAGE. The entire procedure takes two hours - that is, the analysis in this case is much quicker than RNA sequencing in solution. Time saving is the greatest when a large number of RNA samples are analyzed, and solid-phase sequencing allows many RNA fragments to be processed simultaneously.

Polony Sequencing

Polony sequencing, developed by George M. Church at Harvard Medical School, is a sequencing technique that uses paired-tag library emulsion PCR to amplify the target DNA, and sequencing by ligation to detect DNA bases.

George M. Church, looking magnificent.

When polony sequencing was published was released in 2003, and the cost was less than 10% of Sanger Sequencing. It was used to sequence a full E. coli genome in 2005 with an error rate of less than 0.00001%.

One unique aspect of polony sequencing is that its technology is an open-source platform. This means the software and protocols are free and don't require licensing or a fee for use. Any modifications or improvements to the system are also made available. Additionally, the only machinery required is a computer-controlled fluidics system and an epifluorescence microscope.

Procedure

The procedure takes a total of 9 steps:

1. Shearing DNA

The first step, as in any other NGS technique, is the library construction. We break apart the genomic DNA.

2. DNA Repair

Next we want to perform *end-repair* to fix any damaged or incompatible edges. We want to make our DNA ends blunt-ended with a phosphate group attached at the 5'. This allows us to ligate any adapter oligonucleotides.

The DNA fragments also undergo *A-tailed treatment*. This adds an A to the 3' end of the sheared DNA.

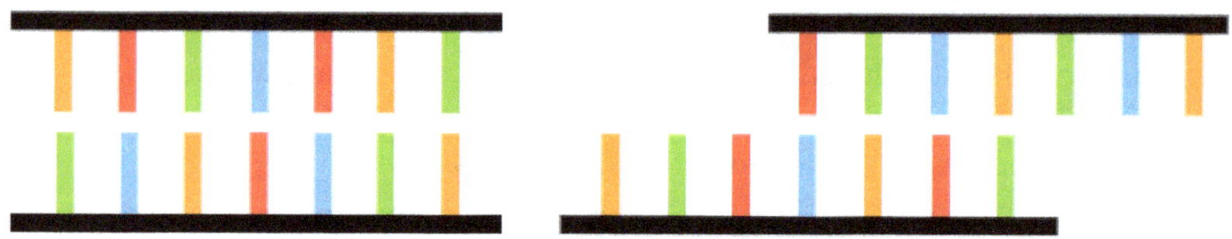

The left dsDNA has blunt ends, while the right has sticky ends.

After the DNA molecules are repaired, those of length 1kb are selected by loading them onto a 6% TBE PAGE gel.

3. DNA Circularization

The next step is to circularize the DNA. We do this with the T-tailed 30 bp long synthetic oligonucleotides (T30). This contains two outward-facing *MmeI* recognition sites.

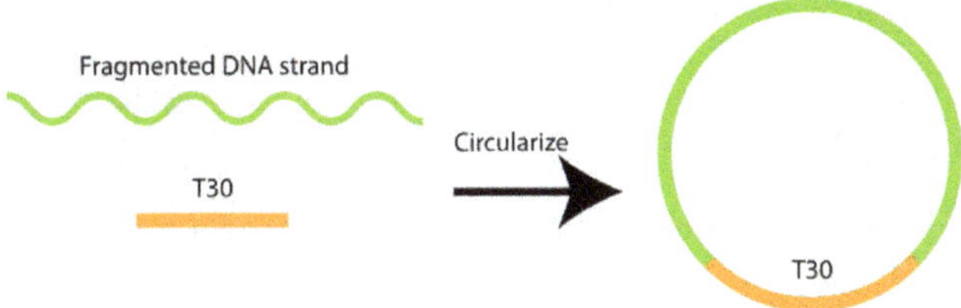

Circularization with T30.

4. Rolling Circle Replication

The circularized DNA undergoes *rolling circle replication*. This is a type of nucleic acid replication that rapidly synthesizes multiple copies of circular molecules of DNA.

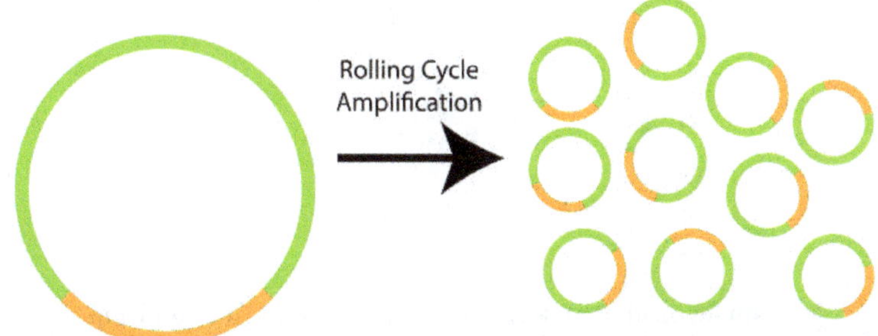

Rolling Cycle Amplification to generate several copies of the circularized DNA.

The newly generated circularized DNA are then digested by restriction enzyme Mmel (type IIs restriction endonucleases), which cut at a distance away from its recognition site. This releases the T30 fragment, flanked by 17-18 bp tags of the sequence (70 bp in total).

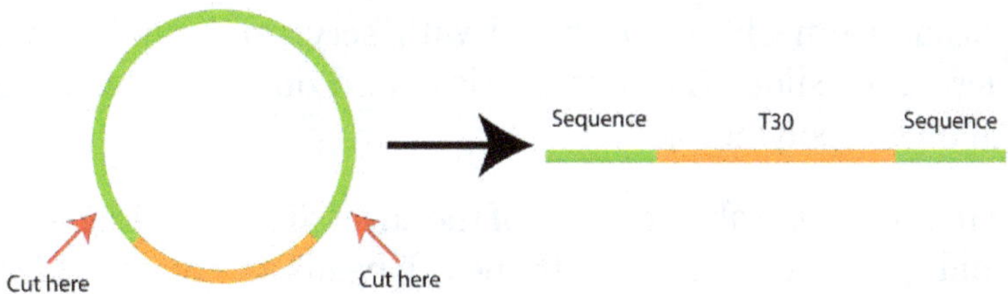

Digest the circularized DNA with Mmel, which cut a specific number of bp away from the recognition site.

5. DNA Repair and Primers Added

The resulting DNA is repaired and FDV2 and RDV2 are added on each ends. In total, this results in a 135 bp library molecules.

We now have DNA templates with 44 bp FDV sequence, a 17-18 bp proximal tag, the T30 sequence, a 17-18 bp distal tag, and a 25 bp RDV sequence.

FDV2	Sequence	T30	Sequence	RDV2
44 bp	17-18 bp		17-18 bp	25 bp

Attach primers FDV2 and RDV2 for emulsion PCR.

6. ePCR

ePCR is used to amplify the 135 bp paired end-tag library molecules. This process takes place within a water droplet embedded within an oil solution. Check out our more thorough explanation on emulsion PCR.

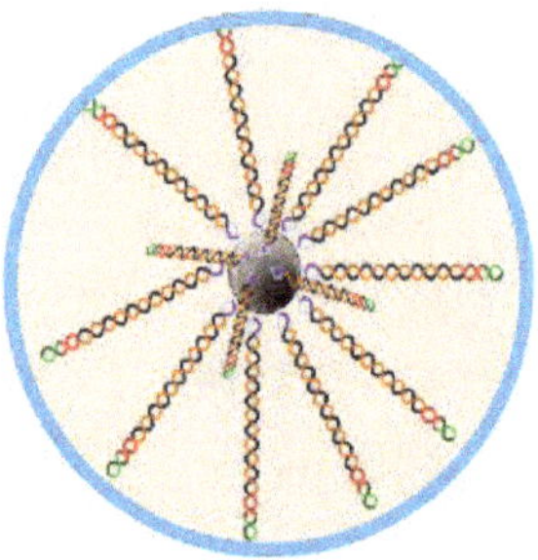

a single bead in an water-in-oil emulsion. After emulsion PCR is performed, thousands of these beads are extracted and placed inside the sequencing platform

7. Coverslip Arraying

Coverslips are washed and treated with *aminosilane*. This eliminates fluorescent contamination and allows for covalent coupling of template DNA and beads to attach.

The resulting beads from ePCR are mixed with acrylamide and poured into a teflon-masked microscope slide. The coverslip is placed on top of the acrylamide gel for 45 minutes to allow for polymerization.

The beads bind to the aminosaline coating of the coverslip, spreading out in a monolayer in an acrylamide gel. The coverslip with the gel, beads and template DNA are inverted. Now beneath this solution is where the sequencing reagents will flow.

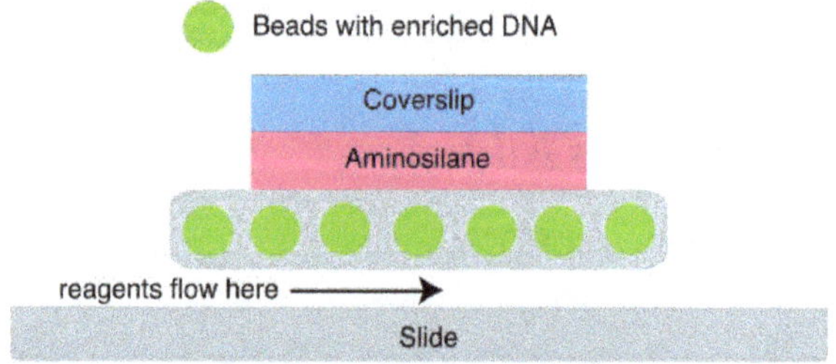

The beads bind to the aminosilane coating to the coverslip and spread out in a monolayer in an acrylamide gel.

8. DNA Sequencing

The methods for DNA sequencing is sequencing by ligation. In short, a series of anchor primers are hybridized to the synthetic oligonucleotide sequences at the genomic DNA sequences.

A group of degenerate nonamers (oligonucleotides of length 9) are used, each with a particularly known query position and fluorescent marker. Thus, in this round the known query is at position 9: Depending on which nonamer binds, we can see which nucleotide is at position 9. We can then do this again to get the nucleotide at position18, then 27, and so on. Now we can use a pool of nonamers that have a known query position down one nucleotide:

We may either use these, or simply shift the known nucleotide position up one base pair and again use nonamers of known query position 1.

We perform throw in this pool of degenerate nonamers again to see nucleotides at positions 8, 16, 24, 32 and so on. We repeat this over again with different known query positions until we are through with the sequence.

Cons

- There may be failures in cleaving the dyes, which can mess up base calls.

Pros

- Cheap: open source, free software.
- Flexible: can include BAC (bacterial artificial chromosome) and bacterial genome re-sequencing, as well as SAGE (series analysis of gene expression) and barcode sequencing.
- Easy to set up: Only need commonly fluorescence microscope, and a computer controlled flow cell.
- Scalable: by using 1 um magnetic bead.

Targeted Analysis Sequencing

Introduction to Targeted Gene Sequencing

Targeted gene sequencing panels are useful tools for analyzing specific mutations in a given sample. Focused panels contain a select set of genes or gene regions that have known or suspected associations with the disease or phenotype under study. Gene panels can be purchased with preselected content or custom designed to include genomic regions of interest.

Next-generation sequencing (NGS) offers the scalability, speed, and resolution to evaluate targeted genes of interest. Multiple genes can be assessed across many samples in parallel, saving time and reducing costs associated with running multiple separate assays. Targeted gene sequencing also produces a smaller, more manageable data set compared to broader approaches such as whole-genome sequencing, making analysis easier.

Advantages of Targeted Gene Sequencing

- Sequences key genes or regions of interest to high depth (500–1000× or higher), allowing identification of rare variants;
- Provides cost-effective findings for studies of disease-related genes;
- Delivers accurate, easy-to-interpret results, identifying variants at low allele frequencies (down to 5%);
- Enables confident identification of causative novel or inherited mutations in a single assay.

Predesigned Targeted Gene Panels

Predesigned panels contain important genes or gene regions associated with a disease or phenotype, selected from publications and expert guidance. By focusing on the genes most likely to be involved, these panels conserve resources and minimize data analysis considerations. Predesigned panels are available for research on various diseases, such as cancer, inherited disorders, cardiac conditions, and autism.

References

- Dna-sequencing: biologydictionary.net, Retrieved 13 July 2018
- Next-generation-sequencing-58: atdbio.com, Retrieved 25 May 2018
- Large-genomes, whole-genome-sequencing, dna-sequencing: illumina.com, Retrieved 24 March 2018
- Exome-sequencing, targeted-resequencing, dna-sequencing: illumina.com, Retrieved 26 July 2018
- Polony-sequencing, ngs-techniques: binf.snipcademy.com, Retrieved 21 June 2018

Chapter 5

DNA Replication

DNA replication is a fundamental process of biological inheritance. It refers to the production of two replicas of DNA from a DNA molecule. This chapter discusses in detail the processes involved in DNA replication, such as polymerase chain reaction and DNA replication in prokaryotic and eukaryotic organisms.

DNA replication is the basis for biological inheritance. It is a fundamental process occurring in all living organisms to copy their DNA. This process is 'semiconservative' in that each strand of the original double-stranded DNA molecule serves as a template for the reproduction of the complementary strand. Hence, the process of DNA replication yields two identical DNA molecules from a single double-stranded molecule. Cellular proof-reading and error-checking mechanisms ensure nearly perfect fidelity of the DNA copies. DNA replication commences at specific locations in the genome called "origins." The DNA unwinds at the origin to form a replication fork.

DNA replication can proceed in only one direction, from the top of the DNA strand to the bottom. Because the strands that form the DNA double helix align in an antiparallel fashion with the top of one strand juxtaposed to the bottom of the other strand, only one strand at each replication fork has the proper orientation (bottom-to-top) to direct the assembly of a new strand in the top-to-bottom direction. For this leading strand, DNA replication proceeds continuously in the direction of the advancing replication fork.

DNA replication cannot proceed along the lagging strand, i.e. the strand with the top-to-bottom orientation, until the replication bubble expands enough to expose a sizeable stretch of DNA. DNA replication then moves away from the advancing replication fork. It can proceed only a short distance along the 'top-to-bottom' oriented strand before the replication process must stop and wait for more of the parent DNA strand to be unwound.

DNA Replication – The Replisome

The replisome is a complex molecular machine that carries out replication of DNA. It is comprised of a number of subcomponents, each performing a specific function during the process of replication. Helicase is an enzyme which breaks the hydrogen bonds between the two strands of DNA, thus separating the strands ahead of DNA synthesis. As helicase unwinds the double helix, it induces the formation of supercoils in other areas of the DNA.

Gyrase relaxes and undoes the supercoiling which has been caused by the helicase by cutting the DNA strands, allowing it to rotate and release the supercoil, and then rejoining the strands. Gyrase is most commonly located up streak of the replication fork, where the supercoils are being formed.

Primase adds complementary RNA primers to a DNA strand to begin Okazaki fragments. In addition, because DNA Polymerae can only continue (but not begin) a strand, primase begins the leading strand as well.

DNA polymerase III is comprised of two catalytic cores- one for replication of the leading strand and one for the lagging strand. DNA polemerase III, however, cannot stay on the DNA strand long enough to efficiently replicate a daughter strand. Hence, DNA polymerase III stays on the strands via a dimer beta clamp which contains three subunits that come together to enclose the strand -- ensuring that DNA polymerase III will remain on the strand for a few thousand nucleotides as opposed to a few hundred.

DNA polymerase I removes the RNA primers set by Primase and completes the Okazaki fragments. Because there is such a small gap remaining after the action of DNA polymerase I has continued the strand of the Okazaki fragment, ligase is required to fill in the gap. The two ends of the Okazaki fragments are subsequently connected by covalent bonds.

Single-strand binding proteins bind to the exposed bases in an effort to counteract their instability and prevent the single-strand DNA from hydrogen-bonding to itself to form dangerous hairpin structures.

DNA polymerases contain a 'proofreading' mechanism, commonly referred to as 'exonuclease activity'. This removes nucleotides that have been mistakenly added.

DNA Replication – Signature of Design

DNA Replication stands as a fundamental challenge to those who seek to hold to a Darwinian worldview. As the process by which biological information is copied and passed on to succeeding generations, the mechanism is fundamental to the process of self-replication of cells. Yet self-replication of cells is necessary for the operation of any selective process such as natural selection. Thus, to attempt to explain the immense sophistication of this mechanism with reference to natural selection requires one to presuppose the existence of the very thing they wish to explain. Because of its extremely sophisticated nature, most biochemists previously reckoned that the system arose once, prior to the origin of the last universal common ancestor. In addition, many biochemists have long regarded the close functional similarity of DNA replication observed in all life as evidence for the single origin of DNA replication. Yet in 1999 researchers from the National Institutes of Health demonstrated that the core enzymes involved in the DNA replication machinery of bacteria and archae/eukaryotes (the two major trunks of the evolutionary tree of life) did not in fact share a common evolutionary origin. It thus appears as if two identical DNA replication systems have emerged independently in bacteria and archae- after these two evolutionary lineages supposedly diverged from the last universal common ancestor.

It is phenomenal to think that this engineering marvel evolved a single time, let alone twice. There exists no obvious reason for DNA replication to take place by a semiconservative, RNA primer-dependent, bidirectional mechanism that depends on leading and lagging strands to produce DNA daughter molecules. Even if DNA replication could have evolved independently on two separate occasions, it is reasonable to expect that fundamentally different mechanisms would emerge for bacteria and the archae/eukaryotes given their idiosyncrasies.

Structure of DNA

Before we jump into the process of replication, let us take a quick look at the structure of DNA. As we all know, DNA is the genetic code that helps our cells to develop and reproduce in a planned way. Because of which it is called the 'Blueprint of Life'. Getting back to its structure, DNA is made up of four nucleotides. Thinking what Nucleotides are? They are molecules, which are made of a phosphate group, a sugar ring, and a nitrogen base. These nucleotides are Adenine (A), Thymine (T), Guanine (G), and Cytosine (C). A and G are called Purines while T and C are called Pyrimidines. Those words can be a mouthful but you will be able to read them after a little bit of practice.

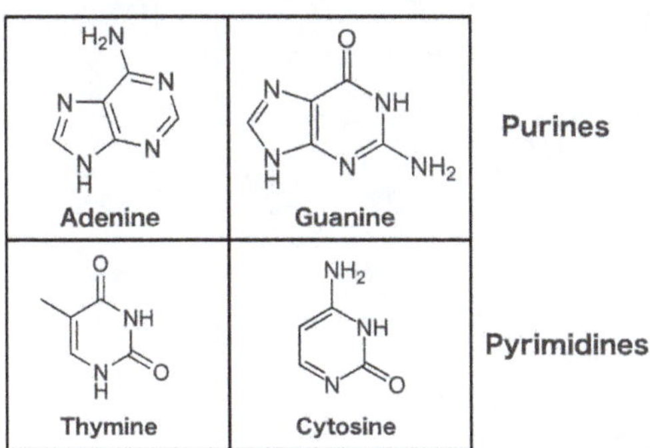

DNA is made of two strands, these strands have nucleotides lined up one after the other and those nucleotides are bound to the nucleotides on the other strand to create a ladder-like structure. Now the binding between nucleotides is very specific and the binding is via Hydrogen Bonds. A will bind to T and C will bind to G. These nucleotides bind to each other and are called as Base pairs. So there we have it. A seemingly never-ending ladder made of nucleotides pairing up with each other. But there is one more change, take that ladder and twist it. That's it, our DNA looks like a simple double helix with specific nucleotide binding.

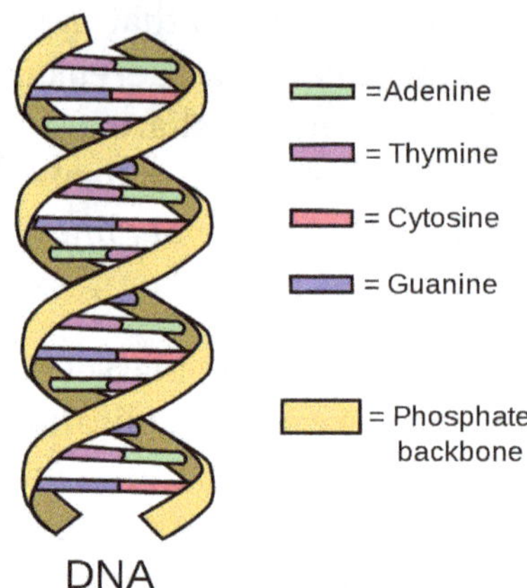

DNA double helix

Directionality

These strands have two designated ends called 5' and 3' (you can read that as 5 prime end and 3 prime end). These numbers indicate end-to-end chemical orientation. The numbers 5 and 3 represent the fifth and third carbon atom of the sugar ring respectively. 5' is the end, which joins a phosphate group that attaches to another nucleotide. 3' end is important as during replication the new nucleotide is added to this end.

In terms of direction, if one strand is 5' to 3' while reading from left to right, the other strand will be 3' to 5'. Simply put, the strands run in opposite directions. This orientation is kept for easy binding between nucleotides of the opposite strands.

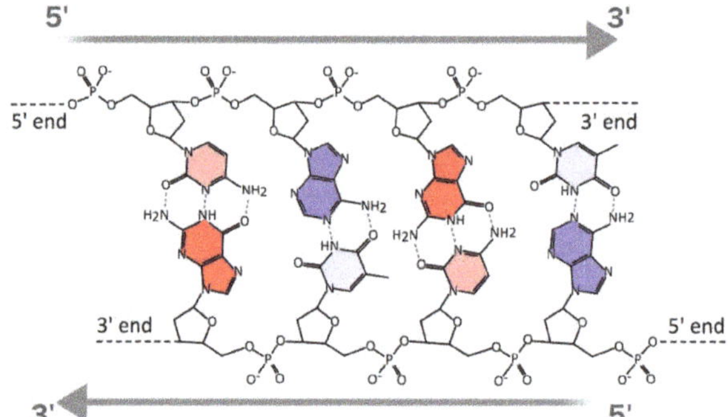

The chemical structure of a four base pair fragment of a DNA double helix

Process of Replication

Replicating the entire DNA is no easy job. The human genome (Genome means a complete set of genes present in the cell) has around 3 billion base pairs (Nucleotide pairing, remember?). So to make a copy of something that long would take a lot of time. But it doesn't. Because our cells have a set of enzymes and proteins which makes this process quick.

Each enzyme and protein has their own specific function. Let us look at the process step by step.

Initiation

- Helicase: The point at which the replication begins is known as the Origin of Replication. Helicase brings about the procedure of strand separation, which leads to the formation of the replication fork. It breaks the hydrogen bond between the base pairs to separate the strand. It uses energy obtained from ATP Hydrolysis to perform the function.
- SSB Protein: Next step is for the single stranded DNA Binding Protein to bind to the single-stranded DNA. Its job is to stop the strands from binding again.
- DNA Primase: Once the strands are separated and ready, replication can be initiated. For this, a primer is required to bind at the Origin. Primers are short sequences of RNA, around 10 nucleotides in length. Primase synthesizes the primers.

Elongation

- DNA Polymerase III: This enzyme makes the new strand by reading the nucleotides on the template strand and specifically adding one nucleotide after the other. If it reads an Adenine (A) on the template, it will only add a Thymine (T). It can only synthesize new strands in the direction of 5' to 3'. It also helps in proofreading and repairing the new strand. Now you might think why does Polymerase keep working along the strand and not randomly float away? It's because a ring-shaped protein called as sliding clamp holds the polymerase into position.

Now when replication fork moves ahead and the Polymerase III start to synthesize the new strand a small problem arises. The two strands run in the opposite directions. This means that when both strands are being synthesized in 5' to 3' direction, one will be moving in the direction of the replication fork while the other will move in the opposite.

The strand, which is synthesized in the same direction as the replication fork, is known as the 'leading' strand. The template for this strand runs in the direction of 3' to 5'. The Polymerase has to attach only once and it can continue its work as the replication fork moves forward. However, for the strand being synthesized in the other direction, which is known as the 'lagging' strand, the polymerase has to synthesize one fragment of DNA. Then as the replication fork moves ahead, it has to come and reattach to the new DNA available and then create the next fragment. These fragments are known as Okazaki fragments.

Termination

- DNA Polymerase I: If you remember, we had added a RNA primer at the Origin to help Polymerase initiate the process. Now as the strand has been made, we need to remove the primer. This is when Polymerase I comes into the picture. It takes the help of RNase H to remove the primer and fill in the gaps.
- DNA ligase: When Polymerase III is adding nucleotides to the lagging strand and creating Okazaki fragments, it at times leaves a gap or two between the fragments. These gaps are filled by ligase. It also closes nicks in double-stranded DNA.

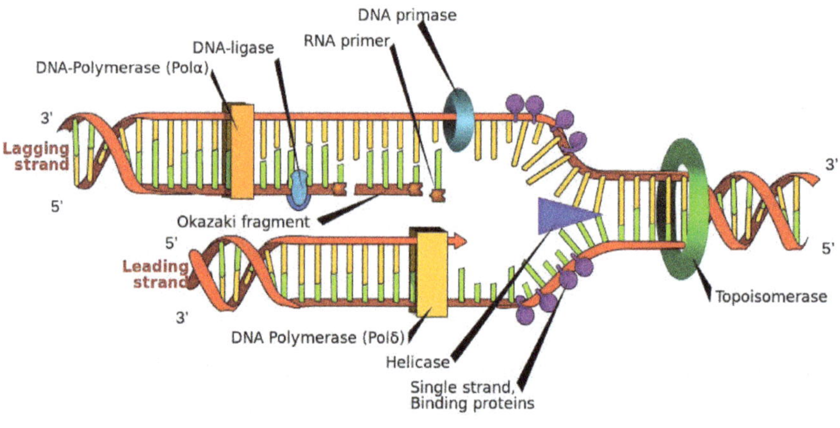

DNA replication

The Replication process is finally complete once all the primers are removed and Ligase has filled in all the remaining gaps. This process gives us two identical sets of genes, which will then be passed on to two daughter cells. Every cell completes the entire process in just one hour. The reason for taking such short amount of time is multiple Origins. The cell initiates the process from a number of points and then the pieces are joined together to create the entire genome.

DNA Replication steps

Step 1: Initiation

The point at which the replication begins is known as the Origin of Replication (oriC). Helicase brings about the procedure of strand separation, which leads to the formation of the replication fork.

Step 2: Elongation

The enzyme DNA Polymerase III makes the new strand by reading the nucleotides on the template strand and specifically adding one nucleotide after the other. If it reads an Adenine (A) on the template, it will only add a Thymine (T).

Step 3: Termination

When Polymerase III is adding nucleotides to the lagging strand and creating Okazaki fragments, it at times leaves a gap or two between the fragments. These gaps are filled by ligase. It also closes nicks in double-stranded DNA.

Prokaryotic DNA Replication

The prokaryotic chromosome is a circular molecule with a less extensive coiling structure than eukaryotic chromosomes. The eukaryotic chromosome is linear and highly coiled around proteins. While there are many similarities in the DNA replication process, these structural differences necessitating some differences in the DNA replication process in these two life forms. DNA replication in prokaryotes has been extensively studied, so we will learn the basic process of prokaryotic DNA replication, then focus on the differences between prokaryotes and eukaryotes.

How does the replication machinery know where to start? It turns out that there are specific nucleotide sequences called origins of replication where replication begins. *E. coli* has a single origin of replication on its one chromosome, as do most prokaryotes shown in figure below. The origin of replication is approximately 245 base pairs long and is rich in AT sequences. This sequence of base pairs is recognized by certain proteins that bind to this site. An enzyme called helicase unwinds the DNA by breaking the hydrogen bonds between the nitrogenous base pairs. ATP hydrolysis is required for this process because it requires energy. As the DNA opens up, Y-shaped structures called replication forks are formed. Two replication forks are formed at the origin of replication and these get extended bi-directionally as replication proceeds. Single-strand binding proteins shown in figure below. Coat the single strands of DNA near the replication fork to prevent the single-stranded DNA from winding back into a double helix.

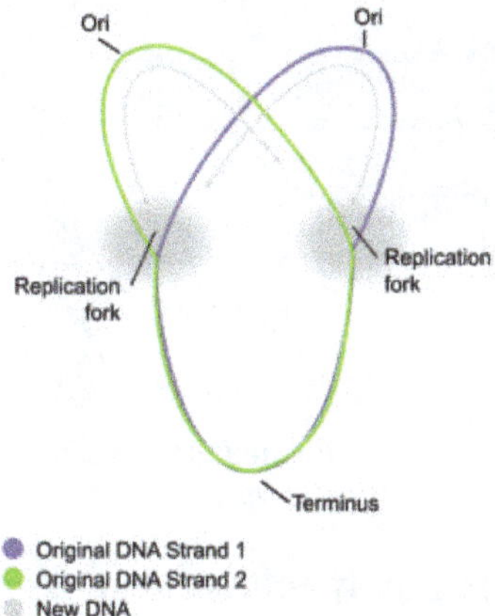

Figure: DNA replication in prokaryotes, which have one circular chromosome.

The next important enzyme is DNA polymerase III, also known as DNA pol III, which adds nucleotides one by one to the growing DNA chain shown in above figure. The addition of nucleotides requires energy; this energy is obtained from the nucleotides that have three phosphates attached to them. ATP structurally is an adenine nucleotide which has three phosphate groups attached; breaking off the third phosphate releases energy. In addition to ATP, there are also TTP, CTP, and GTP. Each of these is made up of the corresponding nucleotide with three phosphates attached. When the bond between the phosphates is broken, the energy released is used to form the phosphodiester bond between the incoming nucleotide and the existing chain.

In prokaryotes, three main types of polymerases are known: DNA pol I, DNA pol II, and DNA pol III. DNA pol III is the enzyme required for DNA synthesis; DNA pol I is used later in the process and DNA pol II is used primarily required for repair (this is another irritating example of naming that was done based on the order of discovery rather than an order that makes sense).

DNA polymerase is able to add nucleotides only in the 5′ to 3′ direction (a new DNA strand can be only extended in this direction). It requires a free 3′-OH group (located on the sugar) to which it can add the next nucleotide by forming a phosphodiester bond between the 3′-OH end and the 5′ phosphate of the next nucleotide. This essentially means that it cannot add nucleotides. If a free 3′-OH group is not available. Then how does it add the first nucleotide? The problem is solved with the help of a primer that provides the free 3′-OH end. Another enzyme, RNA primase, synthesizes an RNA primer that is about five to ten nucleotides long and complementary to the DNA. RNA primase does not require a free 3′-OH group. Because this sequence primes the DNA synthesis, it is appropriately called the primer. DNA polymerase can now extend this RNA primer, adding nucleotides one by one that are complementary to the template strand shown in figure below.

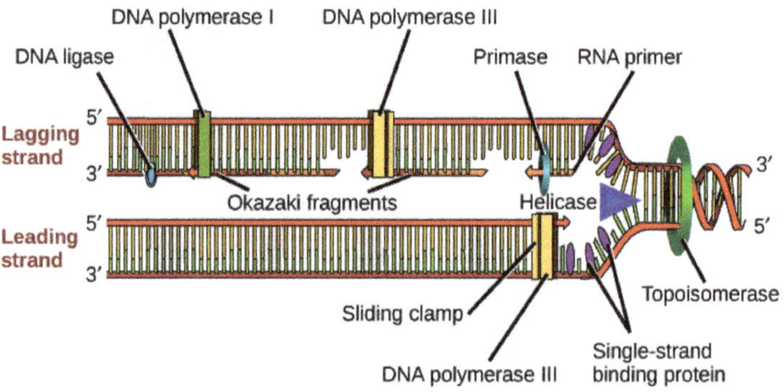

A replication fork is formed when helicase separates the DNA strands at the origin of replication.

The DNA tends to become more highly coiled ahead of the replication fork. Topoisomerase breaks and reforms DNA's phosphate backbone ahead of the replication fork, thereby relieving the pressure that results from this supercoiling. Single-strand binding proteins bind to the single-stranded DNA to prevent the helix from re-forming. Primase synthesizes an RNA primer. DNA polymerase III uses this primer to synthesize the daughter DNA strand. On the leading strand, DNA is synthesized continuously, whereas on the lagging strand, DNA is synthesized in short stretches called Okazaki fragments. DNA polymerase I replaces the RNA primer with DNA. DNA ligase seals the gaps between the Okazaki fragments, joining the fragments into a single DNA molecule.

The replication fork moves at the rate of 1000 nucleotides per second. DNA polymerase can only extend in the 5′ to 3′ direction, which poses a slight problem at the replication fork. As we know, the DNA double helix is anti-parallel; that is, one strand is in the 5′ to 3′ direction and the other is oriented in the 3′ to 5′ direction. One strand, which is complementary to the 3′ to 5′ parental DNA strand, is synthesized continuously towards the replication fork because the polymerase can add nucleotides in this direction. This continuously synthesized strand is known as the leading strand. The other strand, complementary to the 5′ to 3′ parental DNA is extended away from the replication fork, in small fragments known as Okazaki fragments, each requiring a primer to start the synthesis. Okazaki fragments are named after the Japanese scientist who first discovered them. The strand with the Okazaki fragments is known as the lagging strand.

The leading strand can be extended by one primer alone, whereas the lagging strand needs a new primer for each of the short Okazaki fragments. The overall direction of the lagging strand will be 3′ to 5′, and that of the leading strand 5′ to 3′. A protein called the sliding clamp holds the DNA polymerase in place as it continues to add nucleotides. The sliding clamp is a ring-shaped protein that binds to the DNA and holds the polymerase in place. Topoisomerase prevents the over-winding of the DNA double helix ahead of the replication fork as the DNA is opening up; it does so by causing temporary nicks in the DNA helix and then resealing it. As synthesis proceeds, the RNA primers are replaced by DNA pol I, which breaks down the RNA and fills the gaps with DNA nucleotides. The nicks that remain between the newly synthesized DNA (that replaced the RNA primer) and the previously synthesized DNA are sealed by the enzyme DNA ligase that catalyzes the formation of phosphodiester linkage between the 3′-OH end of one nucleotide and the 5′ phosphate end of the other fragment.

Once the chromosome has been completely replicated, the two DNA copies move into two different cells during cell division. The process of DNA replication can be summarized as follows:

1. DNA unwinds at the origin of replication.
2. Helicase opens up the DNA-forming replication forks; these are extended in both directions.
3. Single-strand binding proteins coat the DNA around the replication fork to prevent rewinding of the DNA.
4. Topoisomerase binds at the region ahead of the replication fork to prevent supercoiling (over-winding).

5. Primase synthesizes RNA primers complementary to the DNA strand.
6. DNA polymerase III starts adding nucleotides to the 3′-OH (sugar) end of the primer.
7. Elongation of both the lagging and the leading strand continues.
8. RNA primers are removed and gaps are filled with DNA by DNA pol I.
9. The gaps between the DNA fragments are sealed by DNA ligase.

Table: The enzymes involved in prokaryotic DNA replication and the functions of each.

Prokaryotic DNA Replication: Enzymes and Their Function	
Enzyme/protein	Specific Function
DNA pol I	Exonuclease activity removes RNA primer and replaces with newly synthesized DNA
DNA pol II	Repair function
DNA pol III	Main enzyme that adds nucleotides in the 5′-3′ direction
Helicase	Opens the DNA helix by breaking hydrogen bonds between the nitrogenous bases
Ligase	Seals the gaps between the Okazaki fragments to create one continuous DNA strand
Primase	Synthesizes RNA primers needed to start replication
Sliding Clamp	Helps to hold the DNA polymerase in place when nucleotides are being added
Topoisomerase	Helps relieve the stress on DNA when unwinding by causing breaks and then resealing the DNA
Single-strand binding proteins (SSB)	Binds to single-stranded DNA to avoid DNA rewinding back.

DNA replication has been extremely well-studied in prokaryotes, primarily because of the small size of the genome and large number of variants available. *Escherichia coli* has 4.6 million base pairs in a single circular chromosome, and all of it gets replicated in approximately 42 minutes, starting from a single origin of replication and proceeding around the chromosome in both directions. This means that approximately 1000 nucleotides are added per second. The process is much more rapid than in eukaryotes.

Eukaryotic DNA Replication

The primary effort has focused on replication of the SV 40 (Simian Virus 40) chromosome; these studies have progressed so rapidly in recent years that the SV40 chromosome now can be replicated in-vitro using only eight purified components from mammalian cells. Eukaryotic DNA replication is unidirectional.

- Enzymes involved in DNA Replication;

- DNA Replication: Simple Steps of DNA replication in prokaryotes.

The specific functions of these proteins are highly *reminiscent of proteins* required for replication of plasmids carrying *"Ori.C"*. The DNA molecules in eukaryotic cells are considerably larger than those in bacteria and are organized into complex nucleoprotein structure. There are Eukaryotic DNA Replication origins many choices for appropriate answers.

Mechanism of Eukaryotic DNA Replication

There are three Eukaryotic DNA replication steps required to synthesize new DNA molecule:

Step 1: Initiation

Initiation of Eukaryotic DNA replication is similar to E.Coli, replication, in Eukaryotic replication, it is initiated at a unique location on the SV40 DNA by inter activation of a virus-encoded, site-specific DNA binding protein called *"T.antigen"*.

- Prokaryotic DNA Polymerases;
- Fidelity of DNA Replication.

This multifunctional protein locally unwinds duplex at the SV40 origin also requires ATP and replication factor-A (RF-A). A host cell single-strand binding protein with a function similar to that of SSB Protein in E.Coli cells.

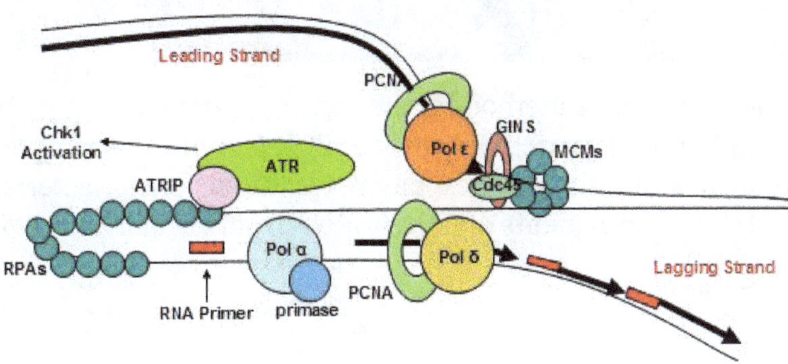

Step 2: Elongation

As in E.coli, eukaryotic DNA replication occurs bidirectionally from RNA primers made by a "Primase" synthesis of the leading strand is continuous, while synthesis of lagging strand is discontinuous.

Two distinct 'Polymerases'- A and d, appear to function at the eukaryotic growing fork. Polymerase d (pol. d) is largely responsible for leading synthesis; polymerase a (pol. a), which is tightly associated with a "Primase", is thought to synthesize the lagging strand.

RNA primers, formed by the action of "Primase", these are elongated for a short stretch by "Pol. a", whose activity is stimulated by replication factor C.

Binding of "PCNA (Proliferating Cell Nuclear Antigen)" at the primer-template terminus then displaces Pol.a, thus interrupting leading-strand synthesis. PCNA increases the "Processivity of the enzyme". The function of PCNA thus appears to be highly analogous to that of the b-subunit of E.Coli polymerase-III.

Step 3: Termination

The termination of replication on linear eukaryotic chromosomes involves the synthesis of special structures called "Telomeres" at ends of the chromosome.

The telomeres consist of repetitive "Oligomeric sequences". The enzyme that prevents this progressive shortening of the lagging strand is a "modified reverse transcriptase" called *"Telomerase"*, which can elongate the lagging-strand template from its 3'-hydroxyl end.

- Deoxyribonucleic acid its Types: A-DNA, B-DNA, and Z-DNA;
- MicroRNA: Special features and its Role;
- Transfer RNA (tRNA).

This unusual enzyme contains a catalytic site that polymerizes deoxyribonucleotides directed by an RNA template as well as the RNA molecule that functions as that template.

Polymerase Chain Reaction

Polymerase chain reaction is method for amplifying particular segments of DNA. It is an enzymatic method and carried out invitro. PCR technique was developed by Kary mullis in 1983. PCR is very simple, inexpensive technique for characterization, analysis and synthesis of specific fragments of DNA or RNA from virtually any living organisms.

Components of PCR

DNA Template

The sample DNA that contains the target sequence. At the beginning of the reaction, high temperature is applied to the original double-stranded DNA molecule to separate the strands from each other.

Dna Polymerase

A type of enzyme that synthesizes new strands of DNA complementary to the target sequence. The first and most commonly used of these enzymes is *Taq*DNA polymerase (from *Thermis aquaticus*), whereas *Pfu*DNA polymerase (from *Pyrococcus furiosus*) is used widely because of its higher fidelity when copying DNA. Although these enzymes are subtly different, they both have two capabilities that make them suitable for PCR: 1) they can generate new strands of DNA using a DNA template and primers, and 2) they are heat resistant.

Primers

Short pieces of single-stranded DNA that are complementary to the target sequence. The polymerase begins synthesizing new DNA from the end of the primer.

Nucleotides (DNTPs or Deoxynucleotide Triphosphates)

Single units of the bases A, T, G, and C, which are essentially "building blocks" for new DNA strands.

RT-PCR

Blocks" for new DNA strands (Reverse Transcription PCR) is PCR preceded with conversion of sample RNA into cDNA with enzyme.

Steps of Polymerase Chain Reaction

DNA replication is a complicated procedure. However, scientists have successfully found a way to carry it out in the controlled environment of a test tube. What goes in that test tube is very important. The scientist adds the DNA or template DNA, followed by a PCR buffer. The PCR buffer is a liquid that maintains optimal pH and salt concentrations, even when other ingredients are added.

With these ingredients ready, we can begin the copying process. Recall that there are three steps in natural DNA replication. During PCR, these are called denaturing, annealing, and extension. Let's look at each step in detail:

1. Denaturation: In order to make a copy of the DNA, the strands must be separated. Instead of using helicase like Mother Nature does, scientists use heat to break the bonds between complementary nucleotides. The DNA is heated to approximately 95 degrees Celsius for 30 seconds. This breaks the bonds and causes the DNA to unwind.
2. Annealing: The next step is to add prime that will serve to flag down the building crew--the polymerase-- so it knows where to start copying. To accomplish this, scientists add PCR primer, a small piece of DNA that dictates which part of the double helix will be copied. During this step, the test tube is lowered to 50 degrees Celsius for 30 seconds.
3. Extension: Now that we've unwound and primed, all that's left is to do is add

on new complementary nucleotides. In nature, DNA polymerase fulfills this duty. But for PCR, scientists use Taq polymerase, a thermo stable polymerase (it comes from an organism found in hot springs, which means it can tolerate the hot temperatures required). Taq polymerase adds the complementary nucleotides to the single-stranded DNA, forming a new, duplicate double helix. This step takes about 60 seconds and requires the test tube temperature to be raised to 72 degrees Celsius.

Factors Affecting PCR

i. Primer

- PCR reaction needs two primer, a forward and a reverse primer.
- Primer is synthesized oligonucleotide usually ranging from 15-30 bases long.
- Primers are designed such that at 3'end they do not have more than two bases complementary to each other as this results in PRIMER-DIMER formation.
- The G + C content are in the range of 40-60%.
- The melting temperature (Tm) of both forward and reverse primer is usually the same.
- Low concentration of primer results in poor yield while high concentration may results in no specific amplification. Hence optimal concentration of primer is needed i.e. 0.1-1μ.

ii. Amount of Template DNA

- Optimal amount of template DNA usually in nano gram. Higher concentration in hibit or results in non-specific amplification.
- Taq DNA polymerase.
- Taq DNA polymerase is 94 KD thermo stable DNA polymerase isolated from Thermus aquaticus.
- Optimal temperature for activity of Taq polymerase is 72° but it can tolerate high temperature and do not affect by denaturating temperature of 94°C.
- Taq DNA polymerase has both 5'-3' polymerase activity and 5'-3' exonuclease activity. But it lacks 3'-5' exonuclease activity (proof reading activity).

Types of PCR

Standard PCR

- Nested PCR

- Random amplified polymorphic DNA
- Long PCR
- Restriction fragment length polymorphism (RFLP)
- Amplified fragment length polymorphism (AFLP)
- Multiplex PCR
- Single cell PCR
- Fast cycling PCR
- In situ PCR
- High fidelity PCR
- Asymmetric PCR
- Repetitive sequence based PCR
- Overlap extension PCR
- Assemble PCR
- Mini primer PCR
- Solid phase PCR
- Touch Down PCR

Reverse Transcriptase Polymerase Chain Reaction (RT-PCT)

- One step RT-PCR
- Two step RT-PCR

Real Time PCR: for DNA or RNA

- Dye binding to ds DNA
- Fluorescent probes

Application

1. Forensic science: DNA finger printing, paternity testing and criminal identification;
2. Diagnosis: Molecular identification of microorganisms;
3. Evolution study: evolutionary biology;
4. Fossil study: paleontology;

5. Gene cloning and expression;
6. Gene sequencing;
7. Vaccine production by recombinant DNA technology;
8. Drug discovery;
9. Mutation study;
10. Human genome project.

Limitations of PCR and RT-PCR

The PCR reaction starts to generate copies of the target sequence exponentially. Only during the exponential phase of the PCR reaction is it possible to extrapolate back to determine the starting quantity of the target sequence contained in the sample. Because of inhibitors of the polymerase reaction found in the sample, reagent limitation, accumulation of pyrophosphate molecules, and self-annealing of the accumulating product, the PCR reaction eventually ceases to amplify target sequence at an exponential rate and a "plateau effect" occurs, making the end point quantification of PCR products unreliable. This is the attribute of PCR that makes Real-Time Quantitative RT-PCR so necessary.

References

- Dna-replication: allaboutscience.org, Retrieved 25 June 2018
- Dna-replication-steps-diagram-where-when-replication-occurs, pure-sciences: scienceabc.com, Retrieved 14 July 2018
- Dna-replication-steps-diagram-where-when-replication-occurs, pure-sciences: scienceabc.com, Retrieved 28 April 2018
- Dna-replication-in-prokaryotes, mhccmajorsbio: openoregon.pressbooks.pub, Retrieved 15 June 2018
- Eukaryotic-dna-replication: golifescience.com, Retrieved 18 June 2018
- Polymerase-chain-reaction-pcr-principle-procedure-steps-types-application: onlinebiologynotes.com, Retrieved 15 April 2018

Chapter 6

Gene Expression

Gene expression is a process, which involves the synthesis of a gene product such as a functional protein or functional RNA from the information derived from a gene. All the diverse aspects of gene expression have been carefully analyzed in this chapter, such as genetic code, antitermination, artificial gene synthesis, gene silencing, RNA splicing, etc.

Gene expression refers to a complex series of processes in which the information encoded in a gene is used to produce a functional product such as a protein that dictates cell function. It involves several different steps through which DNA is converted to an RNA which in turn is converted into a protein or in some cases RNA, for example, genes encoding the necessary information for transfer RNAs and ribosomal RNAs (tRNAs and rRNAs).

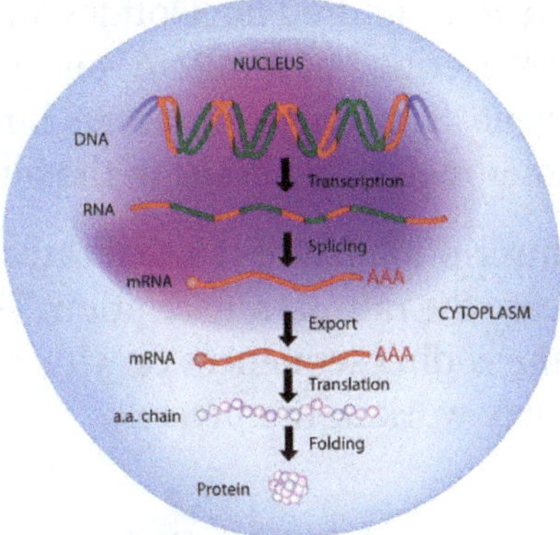

Steps of Gene Expression

The information flow from DNA to RNA to protein can be controlled at several points helping the cell to adjust the quality and quantity of resulting proteins and thus self-regulate its functions. Thus, regulation of gene expression is a critical step in determining what kind of proteins and how much of each protein is expressed in a cell.

Follwing are the two main stages of gene expression.

Transcription

The production of messenger RNA (mRNA) by the enzyme RNA polymerase, and the processing of the resulting mRNA molecule.

Translation: the use of mRNA to direct protein synthesis, and the subsequent post-translational processing of the protein molecule.

Some genes are responsible for the production of other forms of RNA that play a role in translation, including transfer RNA (tRNA) and ribosomal RNA (rRNA).

A structural gene involves a number of different components:

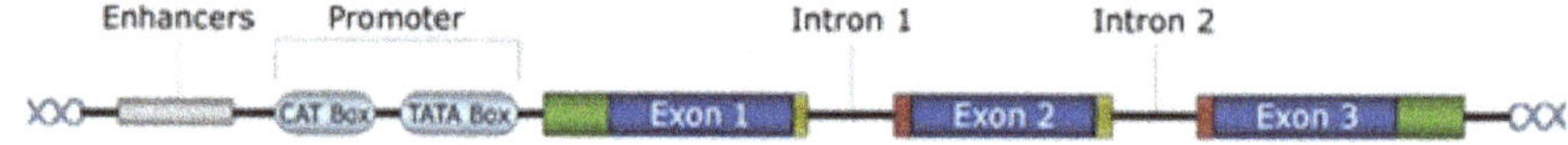

- Exons: Exons code for amino acids and collectively determine the amino acid sequence of the protein product. It is these portions of the gene that are represented in final mature mRNA molecule.
- Introns: Introns are portions of the gene that do not code for amino acids, and are removed (spliced) from the mRNA molecule before translation.

Control Regions

- Start site: A start site for transcription.
- A promoter: A region a few hundred nucleotides 'upstream' of the gene (toward the 5' end). It is not transcribed into mRNA, but plays a role in controlling the transcription of the gene. Transcription factors bind to specific nucleotide sequences in the promoter region and assist in the binding of RNA polymerases.
- Enhancers. Some transcription factors (called activators) bind to regions called 'enhancers' that increase the rate of transcription. These sites may be thousands of nucleotides from the coding sequences or within an intron. Some enhancers are conditional and only work in the presence of other factors as well as transcription factors.
- Silencers. Some transcription factors (called repressors) bind to regions called 'silencers' that depress the rate of transcription.

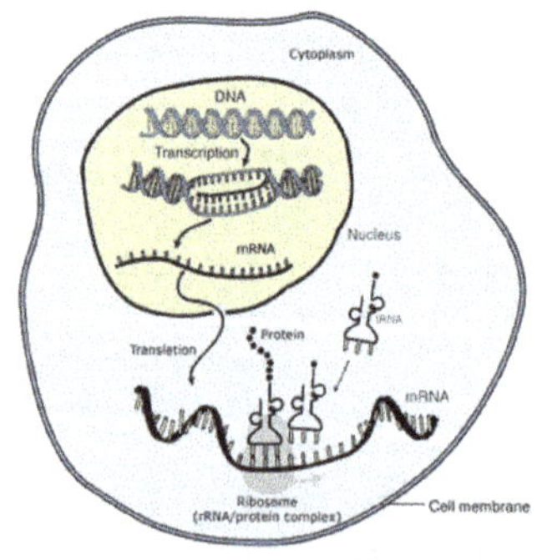

Transcription

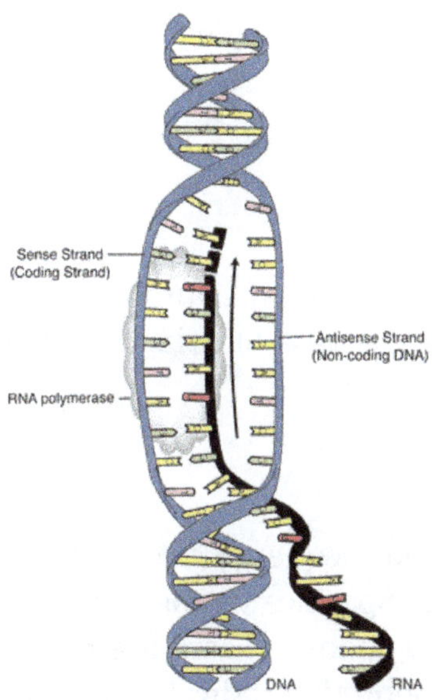

Transcription is the process of RNA synthesis, controlled by the interaction of promoters and enhancers. Several different types of RNA are produced, including messenger RNA (mRNA), which specifies the sequence of amino acids in the protein product, plus transfer RNA (tRNA) and ribosomal RNA (rRNA), which play a role in the translation process.

Transcription involves four steps:

- Initiation: The DNA molecule unwinds and separates to form a small open complex. RNA polymerase binds to the promoter of the template strand (also known as the 'sense strand' or 'coding strand'). The synthesis of RNA proceeds in a 5' to 3' direction, so the template strand must be 3' to 5'.
- Elongation: RNA polymerase moves along the template strand, synthesising an mRNA molecule. In prokaryotes RNA polymerase is a holoenzyme consisting of a number of subunits, including a sigma factor (transcription factor) that recognises the promoter. In eukaryotes there are three RNA polymerases: I, II and III. The process includes a proofreading mechanism.
- Termination: In prokaryotes there are two ways in which transcription is terminated. In ?-dependent termination, a protein? Is responsible for disrupting the complex involving the template strand, RNA polymerase and RNA molecule. Independent termination, a loop forms at the end of the RNA molecule, causing it to detach itself. Termination in eukaryotes is more complicated, involving the addition of additional adenine nucleotides at the 3' of the RNA transcript (a process referred to as polyadenylation).

- Processing: After transcription the RNA molecule is processed in a number of ways: introns are removed and the exons are spliced together to form a mature mRNA molecule consisting of a single protein-coding sequence. RNA synthesis involves the normal base pairing rules, but the base thymine is replaced with the base uracil.

Translation

After transcription (and, in eukaryotes, after processing), an mRNA molecule is ready to direct protein synthesis. The process of using information in an mRNA to build a polypeptide is called translation.

Genetic Code

During translation, the nucleotide sequence of an mRNA is translated into the amino acid sequence of a polypeptide. Specifically, the nucleotides of the mRNA are read in triplets (groups of three) called codons. There are 616161 codons that specify amino acids. One codon is a "start" codon that indicates where to start translation. The start codon specifies the amino acid methionine, so most polypeptides begin with this amino acid. Three other "stop" codons signal the end of a polypeptide. These relationships between codons and amino acids are called the genetic code.

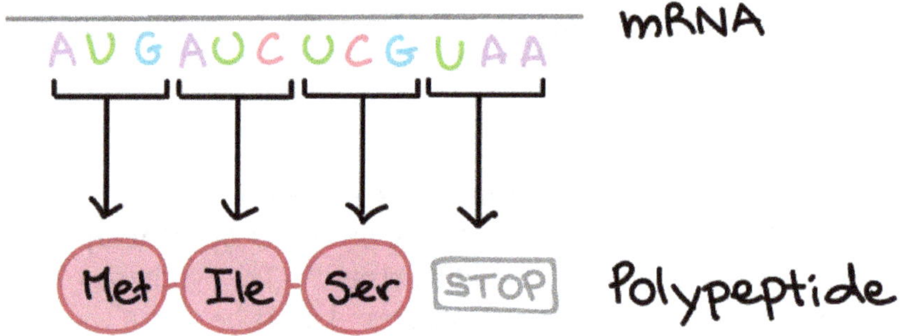

Steps of Translation

Translation takes place inside of structures known as ribosomes. Ribosomes are molecular machines whose job is to build polypeptides. Once a ribosome latches on to an mRNA and finds the "start" codon, it will travel rapidly down the mRNA, one codon at a time. As it goes, it will gradually build a chain of amino acids that exactly mirrors the sequence of codons in the mRNA.

How does the ribosome "know" which amino acid to add for each codon? As it turns out, this matching is not done by the ribosome itself. Instead, it depends on a group of specialized RNA molecules called transfer RNAS (tRNAs). Each tRNA has a three nucleotides sticking out at one end, which can recognize (base-pair with) just one or a few particular codons. At the other end, the tRNA carries an amino acid – specifically, the amino acid that matches those codons.

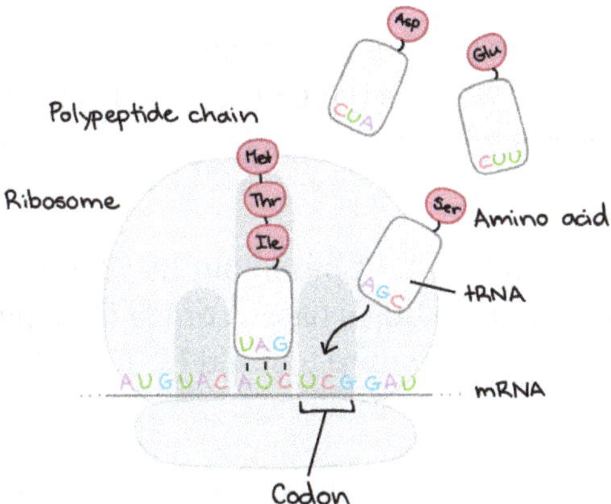

There are many tRNAs floating around in a cell, but only a tRNA that matches (base-pairs with) the codon that's currently being read can bind and deliver its amino acid cargo. Once a tRNA is snugly bound to its matching codon in the ribosome, its amino acid will be added the end of the polypeptide chain.

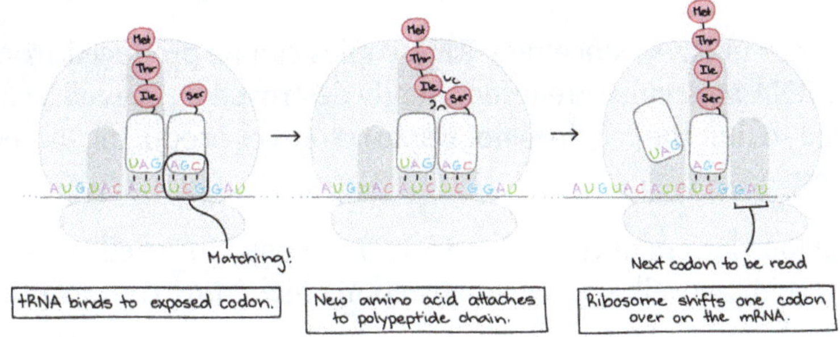

This process repeats many times, with the ribosome moving down the mRNA one codon at a time. A chain of amino acids is built up one by one, with an amino acid sequence that matches the sequence of codons found in the mRNA. Translation ends when the ribosome reaches a stop codon and releases the polypeptide.

GeneCalling

In the field of genomics, an important mRNA transcriptional profiling technique is GeneCalling. It measures the levels of cDNA, which represents the levels of gene expression relative to specific transcripts. The differences in gene expression between healthy tissues and tissues that responds to disease or drug use are examined and comparison of the two made using this technology. This technique can be applied for the study of both plant and animal tissues.

Genetic Code

The genetic code is the code our body uses to convert the instructions contained in our DNA the essential materials of life. It is typically discussed using the "codons" found in mRNA, as mRNA is the messenger that carries information from the DNA to the site of protein synthesis.

Everything in our cells is ultimately built based on the genetic code. Our hereditary information – that is, the information that's passed down from parent to child – is stored in the form of DNA. That DNA is then used to build RNA, proteins, and ultimately cells, tissues, and organs.

Like binary code, DNA uses a chemical language with just a few letters to store information in a very efficient manner. While binary uses only ones and zeroes, DNA has four letters – the four nucleotides Adenine, Cytosine, Guanine, and Thymine/Uracil.

Thymine and Uracil are very similar to each other, except that "Thymine" is slightly more stable and is used in DNA. Uracil is used in RNA, and has all the same properties of Thymine except that it is slightly more prone to mutate.

This doesn't matter in RNA, since new RNA copies can be produced from DNA at any time, and most RNA molecules are intentionally destroyed by the cell a short time after they're produced so that the cell does not waste resources producing unneeded proteins from old RNA molecules.

Together, these four letters of A, C, G, and T/U are used to "spell" coded instructions for each amino acid, as well as other instructions like "start transcription" and "stop transcription."

Instructions for "start," "stop," or for a given amino acid are "read" by the cell in three-letter blocks called "codons." When we talk about "codons," we usually mean codons in mRNA – the "messenger RNA" that is made by copying the information in DNA.

For that reason, we talk about codons made of RNA, which uses Uracil, instead of the original DNA code which uses Thymine.

Each amino acid is represented in our genetic instructions by one or more codons, as seen below:

One of the most remarkable evidences for the common descent of all life on Earth from a single ancestor is the fact that all organisms use the same genetic code to translate DNA into amino acids.

There are a few slight exceptions to be found, but the genetic code is similar enough across organisms that when a gene from a plant or jellyfish is injected into a mammal cell, for example, the mammal cell will read the gene in the same way and build the same product as the original plant or jellyfish.

1st base

2nd base	U		C		A		G		2nd base
U	UUU	Phenylalanine	UCU	Serine	UAU	Tyrosine	UGU	Cysteine	U
	UUC	Phenylalanine	UCC	Serine	UAC	Tyrosine	UGC	Cysteine	C
	UUA	Leucine	UCA	Serine	UAA	Stop	UGA	Stop	A
	UUG	Leucine	UCG	Serine	UAG	Stop	UGG	Tryptophan	G
C	CUU	Leucine	CCU	Proline	CAU	Histidine	CGU	Arginine	U
	CUC	Leucine	CCC	Proline	CAC	Histidine	CGC	Arginine	C
	CUA	Leucine	CCA	Proline	CAA	Glutamine	CGA	Arginine	A
	CUG	Leucine	CCG	Proline	CAG	Glutamine	CGG	Arginine	G
A	AUU	Isoleucine	ACU	Threonine	AAU	Asparagine	AGU	Serine	U
	AUC	Isoleucine	ACC	Threonine	AAC	Asparagine	AGC	Serine	C
	AUA	Isoleucine	ACA	Threonine	AAA	Lysine	AGA	Arginine	A
	AUG	Methionine (Start)	ACG	Threonine	AAG	Lysine	AGG	Arginine	G
G	GUU	Valine	GCU	Alanine	GAU	Aspartic Acid	GGU	Glycine	U
	GUC	Valine	GCC	Alanine	GAC	Aspartic Acid	GGC	Glycine	C
	GUA	Valine	GCA	Alanine	GAA	Glutamic Acid	GGA	Glycine	A
	GUG	Valine	GCG	Alanine	GAG	Glutamic Acid	GGG	Glycine	G

Nonpolar, aliphatic Polar, uncharged Aromatic Positively charged Negatively charged

Function of Genetic Code

The genetic code allows cells to contain a mind-boggling amount of information.

Consider this: A microscopic fertilized egg cell, following the instructions contained in its genetic code, can produce a human or elephant which even has similar personality and behaviors to those of its parents. There is a lot of information in there.

The development of the genetic code was vital because it allowed living things to reliably produce products necessary for their survival – and pass instructions for how to do the same onto the next generation.

When a cell seeks to reproduce, one of the first things it does is make a copy of its DNA. This is the "S" phase of the cell cycle, which stands for the "Synthesis" of a new copy of the cell's DNA.

The information encoded in DNA is preserved by the specific pairing of DNA bases with each other. Adenine will only bond with Thymine, Cytosine with Guanine, etc.

That means that when a cell wants to copy its DNA, all it has to do is part the two strands of the double helix and line up the nucleotides that the bases of the existing DNA "want" to pair with.

This specific base pairing ensures that the new partner strand will contain the same sequence of base pairs – the same "code" – as the old partner strand. Each resulting double helix contains one strand of old DNA paired with one strand of new DNA.

These new double helixes will be inherited by two daughter cells. When it's time for those daughter cells to reproduce, each strand of these new double helices, act as templates for a new double helix.

When the time comes for a cell to "read" the instructions contained within its DNA, it uses the same principle of specific pair bonding. RNA is very similar to DNA, and each RNA base bonds specifically to one DNA base. Uracil binds to Adenine, Cytosine to Guanine, etc.

This means that, just like DNA replication, the information in DNA is accurately transferred to RNA as long as the resulting RNA strand is composed of the bases that bind specifically with the bases in the DNA.

Sometimes, the RNA strand itself can be the end product. Structures made of RNA perform important functions in our selves, including assembling proteins, regulating gene expression, and catalyzing the formation of proteins.

In fact, some scientists think that the first life on Earth might have been composed mainly of RNA. This is because RNA can store information in its base pairs just like DNA, but can also perform some enzymatic and regulatory functions.

In most cases, however, the RNA goes on to be transcribed into a protein. Using the amino acid "building blocks of life," our cells can build almost protein machines for almost any purpose, from muscle fibers to neurotransmitters to digestive enzymes.

In protein transcription, the RNA codons that were transcribed from the DNA are "read" by a ribosome. The ribosome finds the appropriate transfer RNA (tRNA) with "anti-codons" that are complimentary to the codons in the messenger RNA (mRNA) that has been transcribed from the DNA.

Ribosomes catalyze the formation of peptide bonds between the amino acids as they "read" each codon in the mRNA. At the end of the process, you have a string of amino acids as specified by the DNA – that is, a protein.

Other building blocks of life, such as sugars and lipids, are in turn created by proteins. In this way the information contained in the DNA is transformed into all of the materials of life, using the genetic code.

Types of Genetic Mutations

Because the genetic code contains the information to make the stuff of life, errors in an organism's DNA can have catastrophic consequences. Errors can happen during DNA replication if the wrong base pair is added to a DNA strand, if a base is skipped, or if an extra base is added.

Rarely, these errors may actually be helpful – the "mistaken" version of the DNA may work better than the original, or have an entirely new function. In that case, the new version may become more successful, and its carrier may out compete carriers of the old version in the population. This spread of new traits throughout a population is how evolution progresses.

Silent Mutations and Redundant Coding

In some cases, genetic mutations may not have any effect at all on the end product of a protein. This is because most amino acids are connected to more than one codon.

Glycine, for example, is coded for by the codons GGA, GGC, GGG, and GGU. A mutation resulting in the wrong nucleotide being used for the last letter of the glycine codon, then, would make no difference. A codon starting in "GG" would still code for glycine, no matter what letter was used last.

The use of multiple codons for the same amino acid is thought to be a mechanism evolved over time to minimize the chance of a small mutation causing problems for an organism.

Missense Mutation

In a missense mutation, the substitution of one base pair for an incorrect base pair during DNA replication results in the wrong amino acid being used in a protein.

This may have a small effect on an organism, or a large one – depending on how important the amino acid is to the function of its protein, and what protein is effected.

This can be thought of like furniture construction. How bad would it be if you used the wrong piece to bolt a chair leg in place? If you used a screw instead of a nail, the two are probably similar enough that the chair leg would stay on – but if you tried to use, say, a seat cushion to connect the leg to the chair, your chair wouldn't work very well.

Nonsense Mutation

A nonsense mutation occurs when the incorrect base pair is used during DNA replication – but where the resulting codon does not code for an incorrect amino acid.

Instead, this error creates a stop codon or another piece of information that is indecipherable to the cell. As a result, the ribosome stops working on that protein and all subsequent codons are not transcribed.

Nonsense mutations lead to incomplete proteins, which may function very poorly or not at all. Imagine if you stopped assembling a chair halfway through.

Deletion

In a deletion mutation, one or more DNA bases are not copied during DNA replication. Deletion mutations come in a huge range of sizes– a single base pair may be missing, or a large piece of a chromosome may be missing.

Smaller mutations are not always less harmful. The loss of just one or two bases can result in a frame shift mutation that impairs a crucial gene, as discussed under "frame shift mutations" below.

By contrast, larger deletion mutations may be fatal– or may only result in disability, as in Di George Syndrome and other conditions that result from the deletion of part of a chromosome.

The reason for this is that DNA is very much like computer source code – one piece of code might be crucial for the system to turn on at all, while other pieces of code might just ensure that a website looks pretty or loads quickly.

Depending on the function of the piece of code that is deleted or otherwise mutated, a small change can have catastrophic consequences – or a seemingly large corruption of code one can result in a system that is just a bit glitchy.

Insertion

An insertion mutation occurs when one or more nucleotides are erroneously added to a growing DNA strand during DNA replication. On rare occasions, long stretches of DNA may be incorrectly added in the middle of a gene.

Like a missense mutation, the impact of this can vary. The addition of an unnecessary amino acid in a protein may make the protein only slightly less efficient; or it may cripple it.

Consider what would happen to your chair if you added a random piece of wood to it that the instructions did not call for. The results could vary a lot depending on the size, shape, and placement of the extra piece.

Duplication

A duplication mutation occurs when a segment of DNA is accidentally replicated two or more times. Like the other mutations listed above, these may have mild effects – or they may be catastrophic.

To imagine if your chair had two backs, two seats, or eight legs. A small duplication and the chair may still be useable, if a little odd-looking or uncomfortable. But if the chair had, for example, six seats attached to each other, it may rapidly become useless for its intended purpose.

Frame-shift Mutation

A frame-shift mutation is a subtype of insertion, deletion, and duplication mutations. In a frame-shift mutation, one or two amino acids are deleted or inserted – resulting in a shifting of the "frame" which the ribosome uses to tell where one codon stops and the next begins.

This type of error can be especially dangerous because it causes all codons that occur after the error to be misread. Typically, every amino acid added to the protein after the frame-shift mutation is wrong.

Imagine if you were reading a book – but at some point during the writing, a programming error happened such that every subsequent letter shifted one letter later in the alphabet.

This is approximately what happens in a frame-shift mutation.

Antitermination

Antitermination is a process in which Rna Polymerase does not recognize a normal Transcription termination signal at the end of a Gene or Operon during transcription and so transcribes the DNA beyond the terminator. Antitermination provides a means of regulating the Expression of genes.

Processive Antitermination

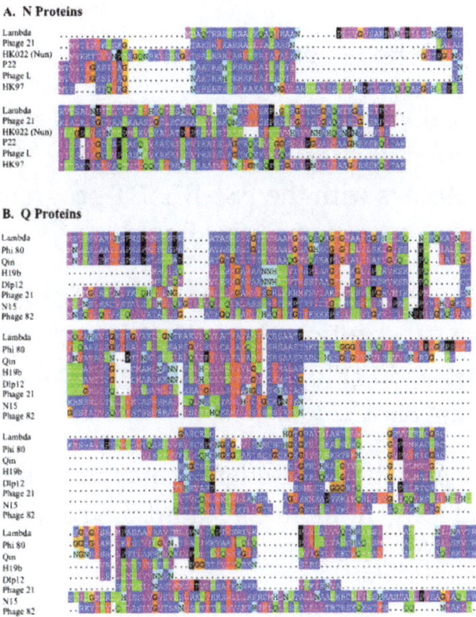

Antitermination in λ is induced by two quite distinct mechanisms. The first is the result of interaction between λ N protein and its targets in the early phage transcripts, and the second is the result of an interaction between the λ Q protein and its target in the late phage promoter. We describe the N mechanism first. Lambda N, A small basic protein of the arginine-rich motif (ARM) family of RNA binding proteins, binds to a 15-nucleotide stem-loop called BOXB. We will capitalize the names of sites in RNA and italicize the names of the corresponding DNA sequences; e.g. BOXB. *Box B* is found twice in the λ chromosome, once in each of the two early operons. It is close to the start point of the P_L operon transcript and just downstream of the first translated gene of the P_R operon. Neither the distance between the transcription start site and *box B*, nor the nature of the promoter (at least in the case of sigma-70-dependent promoters), nor the nature of the terminator is relevant to N action. Although the *box B* sequence is not well conserved in other bacteriophages of the λ family, most of these phages encode proteins that are analogous to λ N and have sequences capable of forming BOXB-like structures in their P_L and P_R operons. In some cases, it has been shown that these structures are recognized by the cognate N analogs. It is believed that this accounts for the phage specificity of N-mediated antitermination.

(A) Alignment of phage N proteins and the HK022 Nun protein. The color groupings reflect the frequency of amino acid substitutions in evolutionarily related protein domains: an amino acid is more likely to be replaced by one in the same color group than by one in a different color group in related proteins. The amino-proximal ARM regions were aligned by eye and according to the structures of the P22 and λ ARMs complex to their cognate *nut* sites and the remainder of the proteins were aligned by Crustal W. The dots indicate gaps introduced to improve the alignment. Aside from the ARM regions, the proteins fall into three very distantly related (or unrelated) families:

(i) λ and phage 21;

(ii) P22, phage L, and HK97;

(iii) HK022 Nun. The divergence of Nun from the N proteins is unsurprising because of their different functions. The sequence database was searched for additional N homologs with the PSI-BLAST program, using each of the listed sequences as a query, but none were found.

Two N proteins were omitted from the alignment: that of phage H19b, because it differs by only three conservative substitutions from N of HK97 (E60D, K80E, and R100K), and that of lambdoid phage φ80 (Phi 80), because it shows no resemblance to any of the other N proteins, lacking an ARM.

(B) Alignment of phage Q proteins. The alignments were generated by Crustal W and the database was searched for Q homologs as described above. These proteins fall into three very distantly related (or unrelated) families:

(i) λ and Qin;

(ii) H19b, Dlp12, and phage 21;

(iii) N15 and phage 82. Qin and Dlp12 are defective lambdoid pro phages of *E. coli*, but it is likely that their Q proteins are active. The Q proteins of phages HK022 and P22 were omitted from the alignment because of their close similarity to that of λ.

A putative and possibly defective Q, encoded by a sequence located upstream of Shiga-like toxin I genes in an E. coli isolate and found by a BLAST search of the translated nucleotide sequence database, was omitted from the alignment because of its close similarity to the Q of phage H19B.

The structures of complexes between the ARMs of λ N (residues 1 to 22) and P22 N and their cognate BOXBs have recently been determined by nuclear magnetic resonance. The two complexes, although similar, show differences that account for the specificity of N-BOXB recognition. Upon binding, λ and P22 ARMs adopt a bent α-helix conformation that packs against the BOXB hairpin through hydrophobic and ionic interactions. Residues in the amino-proximal segments make multiple base, ribose, and phosphate contacts in the 5′-ascending stem of BOXB without disrupting its regular A-form. The two stem-proximal residues of both RNA loops form a sheared G · A base pair which is contacted by R7 of λ N and R19 of P22 N In the λ complex, the fourth residue (G) of the GAAGA loop is extruded and the remaining residues form a GNRA fold similar to the base-stacked GAAA tetra loop reported in a number of important RNA structures. P22 N also creates a GNRA fold, but this is accomplished by extrusion of the third residue (C) of the GACAA loop. In contrast to the extruded G residue in BOXB-λ, which is not close to λ N, the extruded C residue in BOXB-P22 makes contacts with residues in the carboxyl-proximal segment of P22 N. The structure of the λ N complex is stabilized by an important stacking interaction between W18 and the second residue (A) in the BOXB- λ loop. This interaction is not found in the P22 complex. Formation of the GNRA fold is essential for λ N binding, and mutations in loop residues 1, 3, and 5 that prevent tetra loop formation block N binding. The structure of the complex between the BOXB of phage 21 and its cognate N protein must be substantially different than the complexes described above because the BOXB-21 loop (TCTAACCG) cannot be folded into a GNRA tetra loop. However, HK022 Nun protein which recognizes λ BOXB. Probably does so in a way that resembles that of λ. If so, A3, R7, and W18 of λ N, all of which make base contacts in BOXB, probably correspond to S27, R31, and Y42 of Nun.

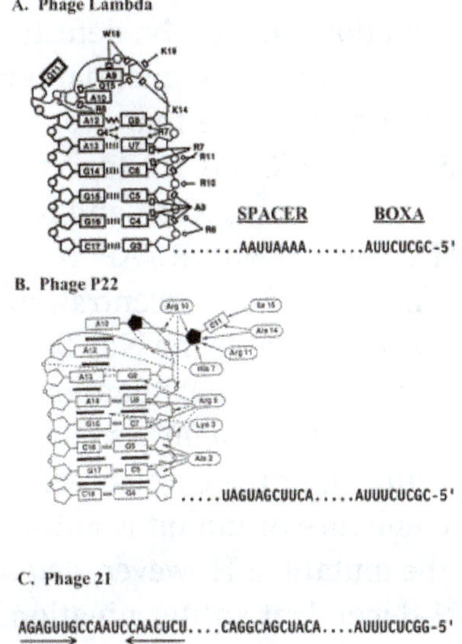

BOXA and BOXB RNAs and their interaction with the ARM of their cognate N proteins

The amino acid-nucleotide interactions are shown to the left except for BOXB of phage 21, for which the structure of the complex is unknown. The sequences of BOXA and BOXA-BOXB spacer are shown to the right. The dots to the left and right of the spacer sequences are for alignment.

(A) λ N-ARM-BOXB complex (adapted from reference with permission of the publisher). Open circles, pentagons, and rectangles represent phosphates, riboses, and bases, respectively. Watson-Crick base pairs (||||) are indicated. The zigzag line denotes a sheared G · A base pair. Open circles, open rectangles, and arrowheads depict ionic, hydrophobic, and hydrogen-bonding interactions, respectively. Guanine-11, indicated by a bold rectangle, is extruded from the BOXB loop.

(B) P22 N-ARM-BOXB complex (adapted from reference with permission of the publisher). Open circles, pentagons, rectangles, and ovals represent phosphates, riboses, bases, and amino acids, respectively. The solid pentagons indicate riboses with a C2′-*endo* pucker. Base stacking, intermolecular hydrogen bonding or electrostatic interactions (<-----), intermolecular hydrophobic or van der Waals interactions (←), intra molecular hydrogen bonds (————) and Watson-Crick base pairs (|||||) are indicated. Cytosine-11 is extruded from the loop. Note that the amino-terminal amino acid residue in the complex corresponds to Asn-14 in the complete protein, and the displayed amino acids are numbered accordingly.

(C) NUTL site of phage 21. The arrows indicate the inverted sequence repeats of BOXB.

Promoter proximal to each of the *box B* sequences is *boxA*, which is also important for antitermination by *λ* N. *Box A*-like sequences are also found in corresponding positions in several other lambdoid phages. Together, the two boxes constitute a *nut* (for N utilization) site, which contains all of the *cis*-acting elements required for N-dependent antitermination. In the presence of N, polymerase becomes termination resistant after transcription of *nut*. Antitermination can still be detected after polymerase has read through thousands of base pairs and many sequential terminators. This implies that the N-dependent modification to polymerase that occurs after transcription of *nut* is stable. BOXA is the loading site for the *E. coli* S10 (or NusE) and NusB proteins. In association with two additional *E. coli* factors, NusA and NusG, an antitermination complex containing N and *λ* NUT is formed with RNAP. It is clear that N is the active factor in the complex, since at sufficiently high concentration, N suppresses transcription termination in vitro in the absence of *nut* or the Nus factors. However, the additional components increase the stability of the antitermination complex and reduce the level of N that is needed. The RNAP α subunit might also have a specific role in N-mediated antitermination. Mutations that alter the carboxyl-terminal domain of the RNAP α subunit have been reported to enhance or inhibit N antitermination in vivo, according to the nature and location of the mutation. However, deletion of the carboxyl-terminal domain of α does not affect N-dependent antitermination in vitro, perhaps indicating that some regulatory component is missing from the reaction.

The formation of the complete antitermination complex can occur in discrete steps. Initially, NusA binds to an N-BOXB complex. This binding requires extrusion of the loop residue of BOXB. Thus, a BOXB tetra loop mutant (GAAGA → GAAA) binds N but does not form an N-BOXB-NusA complex, as demonstrated by super shift experiments. A core complex of BOXB, N, NusA, and RNAP can read through terminators located close to the NUT site in vitro. In the absence of N and NUT, NusA binds to the elongation complex near the 3′-OH terminus of the nascent RNA chain and enhances pausing and termination.

Processive antitermination requires the complete antitermination complex. The assembly of NusB, S10, and NusG onto the core complex involves nt 2 to 7 of λ BOXA (CGCUCUUACACA), as well as the carboxyl-terminal region of N, which interacts with RNAP. The role of NusG in the N antitermination reaction is not clear. NusG binds to termination factor Rho and to RNAP. It stimulates the rate of transcription elongation and is required for the activity of certain Rho-dependent terminators. NusG is a component of the complete antitermination complex and enhances N antitermination in vitro. However, alteration of λ BOXA to a variant called BOXA consensus (CGCUCUUUAACA) allows NusB and S10 to assemble in the absence of NusG. Furthermore, depletion of NusG has no effect on λ N antitermination in vivo, and unlike *nusA,nusB*, and *nusE*, no point mutations in *nusG* that block N activity have been isolated. A NusG homolog, RfaH, enhances elongation of several transcripts in *E. coli* and *S. typhimurium*. The possibility that RfaH and NusG are redundant for N antitermination has not yet been tested, although for several other functions, the two proteins are not interchangeable.

The function of BOXA in λ N-mediated antitermination is likewise not entirely clear. Point mutations in *boxA* that decrease or increase antitermination efficiency have been isolated. On the other hand, deletion of the *boxA* region does not inhibit antitermination in vivo. Instead, antitermination no longer requires NusB. To account for this, it has been proposed that BOXA is not directly required for antitermination but instead is the site of interplay between inhibitory and anti-inhibitory factors. According to this model, *boxA* point mutations that reduce antitermination eliminate binding of the anti-inhibitor but not the inhibitor. *boxA* deletions eliminate binding of the inhibitor, and therefore, the anti-inhibitor, presumably NusB, is no longer needed. This notion is supported by an experiment in which high-level transcription of an antitermination-defective*boxA* point mutant activated growth in *trans* of a λ phage carrying the same mutation in a *nusB* mutant host, presumably by titrating the inhibitor. In a similar experiment, high-level transcription of a consensus BOXA inhibited growth of a phage carrying a wild-type *boxA*, probably by titrating NusB. However, the role of NusB is likely to go beyond that of an anti-inhibitor, and that of BOXA is likely to go beyond that of a site for the interplay of inhibitory and anti-inhibitory factors. In vitro studies with purified proteins show that point mutations in *boxA* impede the assembly of the antitermination complex even in the absence of a known inhibitor and NusB stimulates processive antitermination in such a system. In addition, the role of BOXA in antitermination of Rho-dependent terminators in bacterial rRNA operons appears to be more central than it is in λ, raising the possibility that λ BOXA contributes to antitermination in a way that is at least partially independent of BOXB.

Surprisingly, the λ *nut* sites are also components in a transcription termination pathway. In this pathway, N is replaced by Nun, a protein encoded by a relative of λ, phage HK022. Nun converts antitermination into termination. Other components of the two pathways, notably NusA, NusB, S10, NusG, BOXA, and BOXB, are shared. The sequence similarity of Nun to proteins of the N family, although weak, includes the amino-proximal ARM region. This is unsurprising because Nun, like λ N, binds specifically to BOXB and requires the same BOXB nucleotides for biological function. In vivo, Nun terminates transcription just distal to the *nut* sites. In vitro, Nun arrests RNAP translocation at several positions downstream of *nut*. The arrested elongation complexes contain the 3′ ends of the nascent transcripts in the polymerase active center, and this site remains enzymatically active: the 3′ nucleotide can be removed by pyrophosphorolysis and restored by addition of the appropriate nucleoside triphosphate. However, forward and backward translocation of RNAP is blocked. The Nus factors increase the efficiency of transcription arrest but are not essential if the concentration of Nun is elevated. Nun-dependent release of arrested RNAP from the template and transcript has not been observed in a purified transcription system, presumably because a factor(s) is missing. The differences between N and Nun that lead to their opposed biological activities are unknown. However, the amino-proximal regions, which contain the ARMs, can be interchanged between the two proteins without altering their functions. Therefore, the functional differences are in the carboxyl-terminal 50 to 75% of the proteins. Of particular note is the presence of three C-terminal His residues, specific to Nun. These residues form part of a Zn^{2+} binding motif that is required for Nun activity. The carboxyl-terminal regions of N and Nun may bind different RNAP subunits; certain *rpoC*(β′) mutations block Nun but not N activity.

A second phage-encoded factor, λ Q protein, induces antitermination in the λ late operon. Lambda Q, like λ N, has functional analogs in other phages (Fig. **1**). These late antiterminators probably act by a similar mechanism, although some are only distantly related or unrelated to λ Q. Initially, Q binds to a region within the λ P_R' promoter. Interaction with RNAP can be detected when the transcription complex pauses at +16, downstream of a site similar to the extended −10 sequence of some sigma-70 promoters. The presence of the sigma-70 subunit of RNAP holoenzyme is essential for pausing and for Q-mediated antitermination: RNAP core enzyme that has been artificially paused at +16 by omission of the appropriate nucleoside triphosphate cannot be modified by Q. In addition, sigma-70 mutants that are unable to support Q-mediated antitermination have been isolated. However, once Q has interacted with RNAP holoenzyme, sigma-70 is no longer needed for stable association of Q with the elongation complex. Although antitermination by Q is enhanced by NusA in vitro, it is not clear if the Q reaction has additional requirements in vivo. How Q modifies RNAP function is likewise unknown.

Processive antitermination can be mediated by RNA as well as proteins. Coliphage HK022, alone among the known lambdoid phages, does not encode an analog to λ N. Instead, it promotes antitermination of early phage transcription through the direct action of transcribed sequences called *put* (for polymerase utilization) sites. There are two closely related *put* sites, one located in the P_L operon and the other located in the P_R operon, roughly corresponding to the positions of the *nut* sequences in λ and in other λ relatives.*put* sites act in *cis* to promote read through of downstream terminators in the absence of all HK022 proteins. The *put* transcripts are predicted to form two stem-loops separated by a single unpaired nucleotide. This prediction is supported by mutational studies and the pattern of sensitivity of the two RNAs to cleavage with single- and double-strand-specific endoribonucleases. RNA structure is critical to antitermination because mutations that prevent the formation of base pairs in the stems reduce function, and these mutations can be suppressed by additional mutations that restore base pairing. Like λ N and Q, the PUT sequences suppress polymerase pausing and promote processive antitermination in a purified in vitro transcription system. In contrast to λ N, no phage or auxiliary bacterial factors are required. The only mutations known to block PUT-mediated antitermination change highly conserved amino acids located in a cysteine-rich amino-proximal domain of the RNAP subunit. Strains carrying these mutations are unable to support lytic growth of HK022 but are normal in all other respects tested, including lytic growth of λ and other λ relatives. The phage-restricted phenol types conferred by these mutations suggests that they alter a domain of RNAP-β' that interacts specifically with nascent PUT RNA in the transcription elongation complex, but this idea has not been directly tested. The stability of the putative PUT-RNAP interaction and the nature of the PUT-induced modification to the elongation complex are unknown.

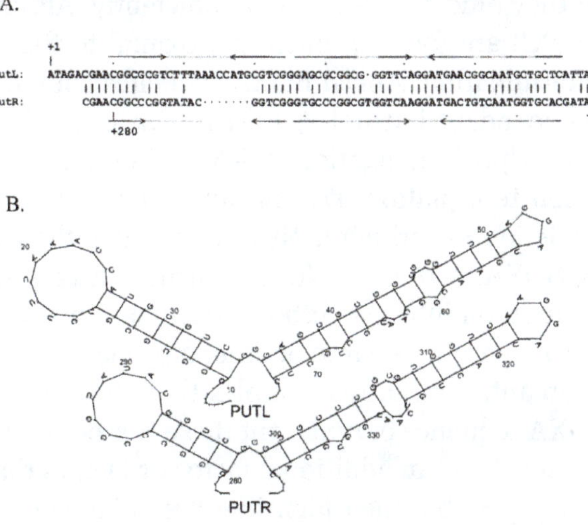

HK022 *put* sites and folded PUT RNAs

(A) Alignment of *putL* and *putR*. The numbers give distances from the start sites of the P_L and P_R promoters, respectively, and the pairs of arrows indicate inverted sequence repeats.

(B) Folded PUTL and PUTR RNAs. The structures, which were generated by energy minimization as described have been partially confirmed by genetic and biochemical studies.

So far, factor-independent antitermination is unique to phage HK022. Both factor-independent and -dependent modes of antitermination are efficient, processive, and well-suited to their tasks. Indeed, the isolation and characterization of a hybrid phage that contains the early promoters and *put* sites of phage HK022 fused to the P_L and P_R operons of λ show that the HK022 antitermination pathway suppresses the λ terminators to the extent necessary for normal lytic growth and lysogen formation by the hybrid. In addition, late gene expression in phage HK022 is activated by a Q-dependent antitermination pathway that is very closely related to that of λ. The relative advantages of the PUT and N-NUT antitermination mechanisms and the evolutionary reasons for their adoption by different closely related phages are obscure. Recent work suggests that the activity of λ N-dependent antitermination is auto regulated so as to couple the lysogeny-lysis decision of infected cells to their growth phase. It is not clear if the *put*-dependent antitermination activity of HK022 is regulated. However, HK022 has devoted the resources saved by the dispensability of N to the production of Nun, a protein that prevents the growth of a potential competitor by coopting a component of its antitermination system.

Processive antitermination was first discovered in a bacteriophage, but examples have since been found in bacterial operons. The *E. coli rrn* operons are regulated by an antitermination mechanism that is dependent on sites that are closely related to λ *boxA* and located promoter proximal to the 16S and 23S structural genes in each operon. The sequences of the *rrn* BOXA sites are more similar to the bacteriophage consensus than is that of λ, and they bind NusB-S10 more efficiently. Although stem-loop structures analogous to BOXB are found promoter proximal to the BOXA sites, they are not essential for antitermination. An*rrn* BOXA sequence confers full antitermination activity against Rho-dependent but not against intrinsic terminators. BOXA also increases the rate of transcription elongation by RNAP. Point mutations in BOXA induce premature transcription termination. *rrn* antitermination requires NusB in vivo, as shown by a NusB depletion experiment. NusA stimulates the elongation rate of *rrn* RNA chains carrying BOXA. A role for NusA is further suggested by the observation that the *nusA10*(Cs) mutation inhibits both antitermination and the rate of transcription elongation in an*rrn* operon. The role of other Nus factors in*rrn* regulation in vivo is not clear. In vitro, an antitermination complex that includes NusA, NusB, S10, and NusG forms at the BOXA sequence of *rrnG*, but these components are not sufficient for antitermination by themselves. An additional factor or factors that can be supplied by a cellular extract are required, but their identities are unknown.

A second bacterial elongation control pathway depends on the RfaH protein, a NusG homolog. RfaH and a *cis*-acting promoter-proximal sequence element, *ops* (for operon polarity suppressor, also called JUMP start), increase the expression of several bacterial operons. The products of these operons affect the production and transport of components located on the outside of the inner membrane, such as lipopolysaccharide core, exopolysaccharide, F pili, and hemolysin. These operons are relatively long, with several genes and intergenic terminators. RfaH and *ops* appear to function together, since mutational inactivation of both elements does not have an additive effect on gene expression. The two elements are thought to act by suppressing termination. First, transcription is increased distal to the promoter but not proximal to a terminator. Second, stimulation of gene expression is not promoter specific. Third, an *rfaH* mutation can be suppressed by a mutation that reduces the activity of transcription termination factor Rho. Purified RfaH stimulates transcription promoter distal to an intrinsic terminator when added to a crude bacterial extract programmed with an*ops*-containing template. However, it has not been shown that the increase results from elongation of transcripts that would otherwise have been terminated. Nevertheless, the requirement for a*cis*-acting site, the ability of the site to act at a distance from terminators, and the homology of RfaH and NusG suggest common elements with the mechanisms of N-dependent antitermination.

Phage P4 has an entirely different mechanism of controlling elongation. It encodes a protein, Psu (for polarity suppressor), that reduces termination by *E. coli* Rho factor. Unlike λ N and Q, Psu does not require *cis*-acting sites to antiterminate and is specific for Rho-dependent terminators. Extracts of cells that contain Psu are deficient in termination at Rho-dependent terminators, and termination can be restored by adding Rho to the extracts. Psu does not act by reducing the level of Rho protein, but it interferes, directly or indirectly, with Rho action. The importance of Rho inactivation in the life cycle of P4 is unclear. Psu stimulates lytic growth of P4, but this is likely to be the result of incorporation of Psu into the P4 capsid rather than (or in addition to) activation of transcription of essential genes that lie downstream of Rho-dependent terminators. The only known protein that is similar to Psu is encoded by a P4 relative, retronphage φR73.

How do terminators and antiterminators act? Do the antitermination pathways described here have common steps? We cannot yet answer these questions, but a brief discussion of what we think we know about the structure and stability of the elongation complex should limit the possibilities and provide a basis for speculation. The active bacterial elongation complex consists of core RNAP, template, and RNA product. The 3′ end of the RNA is engaged in the active site of the enzyme, the following ~8 nt are hybridized to the template strand of the DNA, and the next ~9 nt remain closely associated with RNAP. About 17 nt of the non-template DNA strand are separated from the template strand in the transcription bubble. Elongation complexes can also contain NusA and NusG. These proteins, which increase the stability of the N-mediated antitermination complex, have different effects on elongation. NusA decreases and NusG increases the elongation rate, and both proteins alter termination efficiency in a terminator-specific manner.

An elongation complex, unless located at a terminator, is extraordinarily stable, even when translocation is prevented by removal of substrates. Recent observations suggest that this stability depends mainly on interactions between RNAP and the RNA-DNA hybrid as well as between polymerase and the downstream duplex DNA template. Nascent RNA emerging from the hybrid region and upstream duplex DNA do not appear to be required. The strength of the RNA-DNA hybrid is believed to assure the lateral stability of the complex. Reducing the strength of the RNA-DNA bonds, for example by incorporation of nucleotide analogs, favors backsliding of RNAP on the template, with consequent disengagement of the 3′ RNA end from the active site, and concerted retreat of the RNA-DNA hybrid region from the 3′ end. Such a disengaged complex retains its resistance to dissociation and is capable of resuming elongation if the original or a newly created 3′ end reengages with the active site.

Intrinsic terminators consist of a guanine- and cytosine-rich RNA hairpin stem immediately followed by a short uracil-rich segment within which termination can occur. If termination does not occur at this point, polymerase continues to elongate the transcript with normal processivity until it reaches the next terminator. Neither the stem nor the uracil-rich segment is sufficient for termination, although either can transiently slow elongation. The weakness of base pairing between rU and dA destabilizes the RNA-DNA hybrid in the uracil-rich segment, and this probably contributes to termination. Formation of the hairpin stem as nascent terminator RNA emerges from polymerase destabilizes the RNA-DNA hybrid and interrupts contacts between the emerging nascent RNA and RNAP. It might also interfere with the stabilizing interactions between RNAP and the hybrid or those between RNAP and the downstream region of the template. Cross-linking of nucleic acid to RNAP suggests that both the downstream DNA and the nascent RNA that emerges from the hybrid region, and within which the terminator hairpin might form, are located close to the same regions of the enzyme. Conversely, modifications that render RNAP termination resistant could prevent the terminator stem from destabilizing one or more of these targets, at least while the 3′ end of the RNA is within the uracil-rich segment of the terminator.

The λ N and Q proteins and HK022 PUT RNA also suppress Rho-dependent terminators which, in contrast to intrinsic terminators, lack a precisely determined termination point. Rho is an RNA-dependent ATPase that binds to cytosine-rich, unstructured regions in nascent RNA and acts preferentially to terminate elongation complexes that are paused at nearby downstream sites. Rho possesses RNA-DNA helicase activity, and this activity is directional, unwinding DNA paired to the 3′ end of the RNA molecule. This corresponds to the location of the hybrid and of RNAP in an active ternary elongation complex. The ability of antiterminators to suppress Rho-dependent and -independent terminators suggests that they prevent a step that is common to both classes. Given the helicase activity of Rho, a likely candidate for this step is disruption of the RNA-DNA hybrid. However, other candidates, such as destabilization of RNAP-template or RNAP-hybrid interactions, are also plausible. Alternatively, the ability of N,

Q, and PUT to suppress RNAP pausing, suggests that they prevent Rho-dependent termination by accelerating polymerase away from Rho bound at upstream RNA sites. This explanation raises the problem of why NusG, which also accelerates polymerase, enhances rather than suppresses Rho-dependent termination.

Antitermination Control of Gene Expression

Bacteria have evolved many different complex mechanisms to control both transcription and translation of genes in response to environmental changes. In many cases, transcription is controlled at the level of initiation by DNA-binding proteins that either inhibit (repressors) or stimulate (activators) initiation. In addition, transcription can be regulated at the level of elongation. In some cases, transcription of a gene or operon will terminate prematurely in the absence of the action of a positive regulatory molecule. In these cases, antitermination factors allow transcription to read through termination signals and to generate full-length transcripts.

Two fundamentally different mechanisms for antitermination has been described. In one case, RNA polymerase is modified so as to allow it to read through transcription terminators. This type of mechanism controls phage development and expression of rRNA operons. The second mechanism involves trans-acting factors that interact with RNA and prevent formation of the terminator structure. This mechanism is very similar to attenuation, but antitermination can be distinguished from attenuation in that the action of the regulatory molecule results in transcription read through, with the default pathway being premature termination. In attenuation, the action of the regulatory molecule induces transcription termination, and the default pathway is read through.

Three distinct mechanisms that regulate gene expression by antitermination will be discussed here. These mechanisms differ primarily in the type of biomolecule used as the regulator. The first mechanism uses antiterminator proteins that are activated to bind RNA targets in response to environmental stimuli. In the second mechanism, transfer RNA is used as the regulator. In this case, the degree of aminoacylation (or charging) of the tRNA is used to sense the availability of the cognate amino acid within the cell, to induce expression of genes involved in metabolism of this amino acid. Finally, in the case of the Escherichia coli tryptophanase operon, it appears that ribosomes are used as the regulatory molecule. Thus, bacteria have evolved a large number of mechanisms to use different biomolecules to all perform the same task—that is, to alter the conformation of the nascent mRNA to signal RNA polymerase whether it should terminate prematurely or continue transcription of the particular structural gene.

RNA-Binding Protein-Mediated Antitermination: The Sac/Bgl Family of Antiterminator Proteins

Expression of several catabolic operons in bacteria is regulated by antitermination involving RNA-binding proteins. These proteins prevent formation of Rho-independent transcription terminators in the nascent mRNA upstream of the regulated gene(s). One such system in E. coli and several in Bacillus subtilis, appear to be highly related based on similarities of their antiterminator proteins as well as their RNA targets. In addition, several other systems function similarly but appear to have arisen independently. A general model for this mechanism is shown in figure below.

A general model for antitermination control by the Sac/Bgl family of antiterminator proteins. Under non inducing conditions, transcription starts at the promoter (designated by the arrow) and terminates prematurely, often in a leader region prior to the structural genes. In the presence of inducer, the antiterminator protein is activated to bind to the RAT (ribonucleic antiterminator) RNA. This binding stabilizes an RNA secondary structure involving the RAT, which prevents formation of the overlapping terminator, and transcription continues into the structural genes.

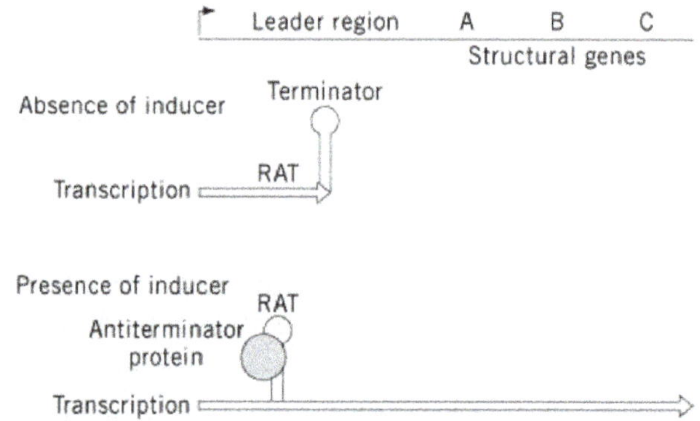

The E. Coli Bgl Operon

The E. coli bglGFB operon: Encodes all the functions necessary for the regulated uptake and utilization of aromatic b-glucosides. The operon is cryptic in wild-type strains but can become functional through spontaneous mutations. When functional, expression of this operon is regulated by antitermination mediated by the BglG protein in response to the levels of b-glucosides. In the absence of inducer, BglG does not bind RNA, and most transcripts terminate at one of two Rho-independent transcription terminators present in the leader region upstream of bglG and between bglG and bglF. When b-glucoside levels are high, BglG binds to an RNA target, named RAT for ribonucleic antiterminator, just upstream of the terminators. This binding stabilizes an alternative antiterminator RNA structure, which prevents formation of the terminator, thus allowing transcription to continue and the operon to be expressed.

The RNA-binding activity of BglG is: Regulated by phosphorylation mediated by BglF. In the absence of b-glucosides, BglF phosphorylates BglG, which prevents it from dimerizing and binding to the RAT. In the presence of b-glucosides, BglF dephosphorylates BglG, which now dimerizes and binds to the RAT. Phosphorylation of both b-glucosides and BglG is accomplished by transfer of the phosphate group from the same phosphorylated residue, Cys24, in BglF. These results suggest that, under conditions in which b-glucoside levels are high, the phosphate group can be transferred from BglG back to Cys24 in BglF. A model has been proposed in which unliganded BglF phosphorylates BglG, and b-glucoside binding induces BglF to undergo a conformational change that activates it to dephosphorylate BglG.

A similar system for b-glucoside utilization: Exists in the related Gram-negative enteric bacterium Erwinia chrystanthemi, although in this case the arb operon is not cryptic. ArbG shows high sequence similarity to BglG, suggesting that it functions analogously in antitermination control of the E. chrystanthemi arb genes. Antitermination also appears to control b-glucoside operons in several Gram-Positive Bacteria as well. A putative b-glucoside (bgl) operon has also been identified in B. subtilis and may be regulated by a similar antitermination mechanism. In addition, a protein, BglR, with homology to BglG also controls b-glucoside usage in Lactococcus lactis.

The B. Subtilis Sac Genes

Expression of two sucrose utilization operons in B. subtilis, sacPA and sacB, is induced by sucrose via transcription antitermination mediated by the RNA-binding proteins SacT and SacY, respectively. SacT and SacY show extensive sequence similarity to each other, as well as to BglG from E. coli. The antitermination mechanisms that control these genes also appear to be quite similar to that described above for the E. coli bgl operon. Rho-independent transcription terminators exist in leader regions upstream of both sacPA and sacB and prevent transcription of the structural genes in the absence of the inducer, which is sucrose. In the presence of sucrose, SacT and SacY are activated to bind RAT sequences in the sacPA and sacB leader transcripts, respectively, and allow transcription to read through into the structural genes. Like BglG in E. coli, both of these antiterminator proteins are phosphorylated. In the case of SacY, phosphorylation negatively regulates RNA-binding activity and appears to be mediated by SacX. SacT is phosphorylated by HPr, a component of the phosphoenolpyruvate phosphotransferase system, but the role of this phosphorylation in sucrose-mediated antitermination is less clear.

Recently, the protein structure of the RNA-binding domain of SacY has been determined by both NMR and X-ray crystallography. The domain exists as a dimer, with each monomer consisting of a four-stranded antiparallel beta-sheet. Several amino acid residues have been identified through genetic, biochemical, and preliminary NMR studies as being important for RNA binding. These residues are clustered on the surface of one side of the protein structure.

Other Examples of Bgl/Sac Type Antiterminators

In addition to the bgl and sac systems described above, several other operons are regulated by RNA-binding antiterminator proteins with homology to BglG, SacY, and SacT. LicT regulates the licS gene, which is involved in b-glucan utilization in B. subtilis. There is also a RAT sequence overlapping a potential Rho-independent terminator upstream of licS.

In Lactobacillus casei, the lactose (lac) operon is regulated in response to lactose levels by LacT, which shows sequence homology to the other members of the Bgl/Sac family of antiterminators. The 5′ -leader region of the lac mRNA contains a region with sequence similarity to the RAT sequence, as well as a potential stem-loop structure resembling a Rho-independent terminator.

Antiterminators with No Similarity to the Bgl/Sac Family

Several other systems are regulated by RNA-binding antiterminator proteins that are unrelated to those of the Bgl/Sac family; furthermore, these proteins do not appear to be related to each other. These regulatory systems thus appear to have arisen independently.

In B. subtilis, both the glp regulon, which is involved in usage of glycerol-3-phosphate, and a histidine-utilization (hut) operon are regulated by RNA-binding antiterminator proteins; GlpP and HutP, respectively. The amino acid sequences of these antiterminator proteins are not similar to any other antiterminator proteins. Further, the mechanisms by which these antiterminator proteins function appear to be different from those described above, because there are no clear antiterminator RNA secondary structures near the terminators in these operons.

The amidase (ami) operon of Pseudomonas aeruginosa is regulated by antitermination in response to short-chain aliphatic amides, such as acetamide. The amiR gene encodes an antiterminator protein (AmiR), which is negatively regulated by AmiC, apparently through formation of an AmiC-AmiR complex. Acetamide destabilizes the AmiC-AmiR complex, leading to antitermination and expression of the operon. AmiR interacts with an RNA target in the 5-leader region of the ami mRNA that contains a Rho-independent terminator. However, no clear antiterminator RNA secondary structure is predicted. AmiR binding has been suggested to function in antitermination by interfering directly with formation of the terminator stem-loop structure.

In addition to all the catabolic operons described above, one anabolic operon has been shown to be regulated by antitermination. Expression of the nas operon of Klebsiella pneumoniae, which encodes enzymes required for nitrate assimilation in this bacterium, is induced by nitrate or nitrite. The NasR protein mediates transcription antitermination through a terminator in the leader region of the operon. This protein shows weak homology with AmiR in the carboxyl-terminal region.

Transfer RNA-Mediated Antitermination

An interesting variation on the antitermination mechanism involves the use of tRNA as the regulatory molecule. This mechanism regulates a large number of aminoacyl-tRNA synthetase genes in Gram-Positive Bacteria and several amino acid biosynthetic operons, including the ilv-leu, in B. subtilis, and the his trp operons in Lactococcus lactis. Expression of these genes is induced specifically by starvation for the corresponding amino acid. In the case of the amino acid operons, insufficient levels of the amino acid leads to increased expression of the corresponding biosynthetic operon. For the amino-acyl-tRNA synthetase genes, increasing the level of the synthetase is thought to allow more efficient charging of the cognate tRNA when the corresponding amino acid pool is low.

A long (approx. 300-nucleotide) untranslated leader region exists upstream of the structural gene(s) of these operons that contains several conserved features, including three stem-loop structures preceding a Rho-independent transcription terminator. Hence, in the absence of the inducing signal, transcription terminates prematurely in the leader region prior to the coding sequences. In addition to the conserved secondary structures, there is an important conserved 14-nucleotide sequence known as the T-box present in each leader region; hence these genes are known as the T-box family. An alternate arrangement of the leader region involving base-pairing between a portion of the T-box and a conserved sequence in the 5′ side of the terminator stem has been proposed to form an antiterminator structure that allows transcription to read through into the structural genes.

Model for ant termination control by tRNA. Under conditions with adequate levels of the cognate amino acid (aa), the charged tRNA does not interact with the leader region, and the terminator forms. Under conditions of starvation for the appropriate amino acid, the uncharged tRNA interacts with the leader region via base-pairing between the anticodon and the specifier sequence, and by base-pairing between the CCA sequence at the acceptor end of the tRNA with the side bulge of the antiterminator in the leader. These interactions stabilize formation of the antiterminator conformation of the leader transcript, resulting in induction of expression of the gene. The tRNA is shown as the shaded cloverleaf structure, and a boxed “A.A.” attached to the tRNA indicates it is aminoacylated.

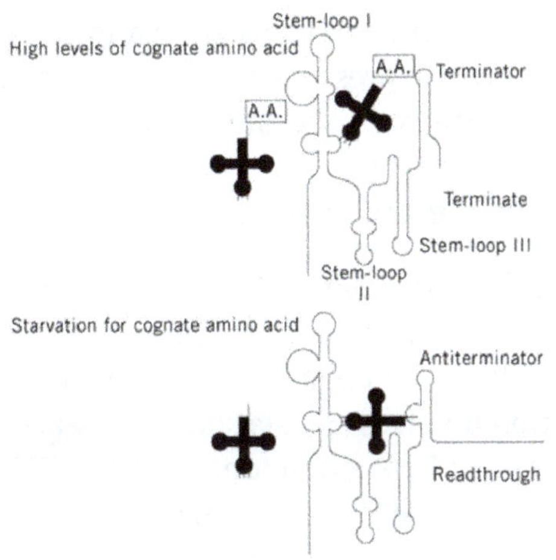

Another important conserved feature: The leader region of these genes is the presence of a triplet sequence corresponding to a codon for the appropriate amino acid for each operon. For example, in tyrS, which encodes tyrosyl-tRNA synthetase, the leader contains a UAC tyrosine codon, while the ilv-leu operon leader contains a CUC leucine codon. This triplet is always present in a bulged sequence in Stem-loop and has been shown to be critical for induction in several systems. It was the presence of these triplets that led to the hypothesis that tRNAs play a role in this regulatory mechanism. This triplet was designated the "specifier sequence" because, in the case of the B. subtilis tyrS gene, altering the sequence to correspond to a codon for another amino acid switched induction to respond to starvation for the new amino acid. Other experiments demonstrated that translation of this codon was not involved in induction and that uncharged tRNA was the inducer. In addition, a second interaction between the CCA sequence at the 3′ end of the uncharged tRNA and the complementary UGG sequence in the T-box have been shown to be important.

A model for tRNA-regulated antitermination: Under starvation conditions for the corresponding amino acid, the cognate uncharged tRNA interacts with two sites in the leader region, to induce formation of the antiterminator structure and allow transcription to read through into the coding region. Aminoacylation of this tRNA is predicted to interfere with the interaction at the CCA end and prevent the charged tRNA from binding; the leader transcript then folds into the conformation with the terminator, halting transcription. It is not known if factors in addition to tRNA are required for antitermination. To date, however, it has not been possible to reconstitute tRNA-mediated antitermination in vitro, and several other lines of evidence also suggest that other factors may be involved in this mechanism.

In addition to the antitermination mechanism: Described above, processing of the leader RNA has been shown to play a role in regulating expression of the B. subtilis thrS gene. Cleavage occurs in the loop of the antiterminator near the T-box sequence and is more efficient under threonine starvation conditions, suggesting that bound tRNA induces both antitermination and RNA processing. This processing increases the stability of the mRNA, which would allow for increased translation and production of the threonyl-tRNA synthetase. Thus induction of expression of this gene in response to threonine starvation occurs at both the level of transcription antitermination and mRNA stability.

The E. coli Tryptophanase Operon

E. coli and several other microorganisms have the capacity to degrade tryptophan as a source of carbon, nitrogen, and/or energy. The degradative tryptophanase operon (tnaCAB) of E. coli is regulated by catabolite repression and by an antitermination mechanism. Antitermination involves translation of a cis-acting 24-residue leader peptide (tnaC) containing a critical Trp codon, one or more RNA polymerase pause sites between tnaC and tnaA, and Rho termination facto. While the precise antitermination mechanism responsible for controlling the tna operon is not firmly established, all of the data are consistent with the following model. During growth in a medium lacking both tryptophan and a catabolite-repressing carbon source, transcription initiation is efficient. As transcription proceeds, translation of the leader peptide occurs as soon as the coding sequence becomes available. Once the translating ribosome reaches the UGA stop codon, ribosome release exposes a rut (Rho utilization) site that immediately follows the stop codon. Rho then binds to the rut site and begins to translocate in the 3′-direction, until it encounters paused RNA polymerase, ultimately leading to transcription termination upstream of tnaA. When cells are growing with inducing levels of tryptophan, TnaC, or a complex of TnaC with an unidentified protein, prevents ribosome release at the tnaC stop codon, thereby masking the rut site and, hence, blocking Rho interaction with the transcript. Eventually RNA polymerase would overcome the pause signal and transcribe the structural genes encoding tryptophanase (tnaA) and a tryptophan permease (tnaB). This model assumes that there is a fundamental difference between the TnaC peptide, the TnaC peptide-protein complex, in cells growing with or without tryptophan. It was proposed that such a complex under inducing conditions would prevent ribosome release, reminiscent of characterized translation attenuation mechanisms. The tryptophanase operon of Proteus vulgaris is thought to be regulated by a mechanism essentially identical to that of E. coli.

Artificial Gene Synthesis

Artificial gene synthesis is the chemical synthesis of a DNA sequence that represents one or more genes. While site-directed mutagenesis is regularly used to probe biological hypotheses by incorporating single base mutations, artificial gene synthesis provides a method to efficiently produce long stretches of natural and non-natural nucleic acid sequences, broadening the scope of biological experiments. Sequences that are hard to isolate from natural sources can be routinely generated in the lab, even entirely non-natural gene sequences can be synthesised. Not only does gene synthesis have implications for synthetic biology, it also introduces the possibility of developing genes containing modified nucleotides or even novel base pairs that could allow for the expansion of the genetic code.

An initial challenge of gene synthesis was the production of long stretches of nucleotide sequence: genes can range from several hundred to several thousand base pairs in length. Current solid-phase oligonucleotide synthesis technology is limited to the generation of oligos 200 nucleotides in length. Two major methods exist to circumvent this issue: polymerase cycling assembly and synthesis by ligation.

Polymerase Cycling Assembly (PCA)

Polymerase cycling assembly (PCA) exploits single stranded templates rather than relying on total synthesis of both DNA strands. Short oligos are synthesised that alternate between both strands of the gene, overlapping by regions of 20-30 base pairs. Then, to give a superficial explanation, a polymerase uses dNTPs to "fill in the gaps" on one gene strand using the complementary bases on the other. A subsequent, entirely separate polymerase step then amplifies the constructed gene by traditional polymerase chain reaction (PCR).

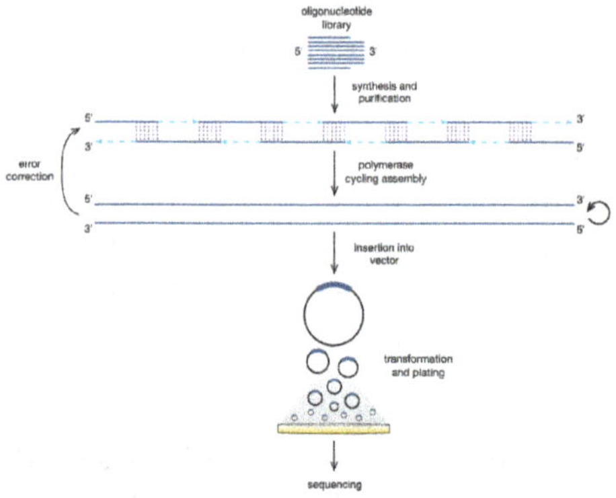

Gene synthesis by polymerase cycling assembly (PCA).

With PCA, the commonly used explanation of the polymerase "filling in the gaps" is misleading: the gene is actually assembled step by step. Overlapping oligos anneal at their termini to form discrete overhanging dimers. The overhangs act as templates to elongate the complementary oligo by polymerase action generating fully double-stranded DNA that makes up a small section of the gene. These extended dsDNA sections are then melted and annealed to another overlapping oligo further along the gene, forming more single stranded overhangs. Further polymerase action turns these dimers into longer dsDNA fragments and this process continues until the entire gene is assembled.

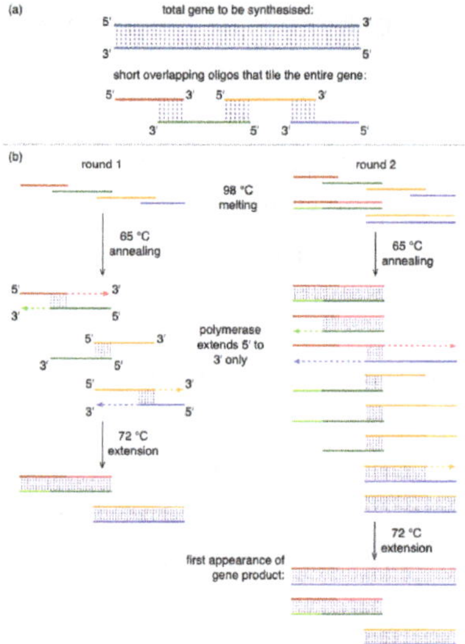

Gene synthesis by PCA – 4 oligos.

As shown in Figure above, many combinations of dimers can anneal and not all result in successful extension. For examples of any more than 4 oligos, the various iterations of annealing and elongation get too complicated to represent diagrammatically. One consistent point however is that polymerases extend 5' to 3' only, so although many oligo dimers will overlap and anneal, only those with the potential to extend 5' to 3' will be elongated. This limitation implies that the terminal oligos that cover the ends of the gene must be on the gene strand with the 5' terminus. Otherwise, if a terminal oligo were on the strand with the 3' terminus, the complementary oligo would have to be extended 3' to 5' to complete the gene, which is impractical for standard polymerase methodologies.

PCA eliminates the need for the whole gene to be synthesised solid phase. Once assembled, the gene product can be amplified by traditional PCR to generate a suitable concentration and purified by filtration to remove any remaining primers.

Error Correction

Unfortunately, the process is not without errors. Insertion and deletion mutations (frameshift mutations) can be introduced at the solid phase synthesis stage and chain extension by polymerase risks further substitutions, insertions and deletions. Fortunately, natural DNA repair enzymes exist to correct these errors. However, since synthetic DNA has no indication of which strand has the correct sequence nucleases that rely on positive identification of the correct template strand cannot be used. Instead, it is necessary to use endonucleases that completely excise stretches of DNA that contain a mutation.

Since PCR cannot discriminate between correct sequences and those with errors, any mutations that have accumulated in the solid phase synthesis or polymerase extension of one gene strand will be matched in the reverse complement during amplification: the mutation is "invisible". Therefore, the mixture of correct and incorrect gene products is melted and reannealed. It is statistically unlikely that mutations will occur in the same place in the same way on two random strands. As such, random reannealing to a strand which likely doesn't have the complementary mutation generates a mismatch as shown in figure below. This mismatch perturbs the secondary structure of the gene.

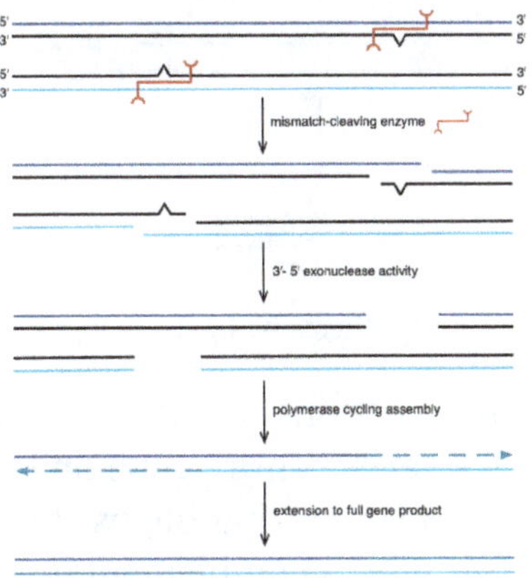

Error correctionRandom errors present in dsDNA sequence will present as mismatches.

Certain mismatch repair enzymes such as T7 Endonuclease I recognise these structural perturbations and cleave the first, second or third phosphodiester bond 3' to the mismatch (dependent on the specific enzyme) on both strands. A single strand exonuclease then chews up the overhanging single stranded DNA 3' to 5'. This process produces dsDNA of varying sizes that all overlap and between them contain the correct sequence spanning the entire gene. Subsequent PCA and PCR amplification generates the corrected gene product.

Gene Synthesis by Ligation

While PCA facilitates the synthesis of large genes and genomes, even after error correction the product often has many errors present in the sequence. Gene synthesis by ligationhas a lower error rate than PCA, but is limited to the synthesis of smaller genes since the fragment oligos must span the entire sequence of both the sense and antisense strands: this is impractical for longer genes.

However, if the goal is to synthesise a smaller gene (< 2 kbp), gene synthesis by ligation will likely generate a product with fewer errors because:

- It removes the need for a polymerase that can often introduce mutations;
- Strands are subject to hybridisation selection – since all oligos overlap entirely, errors in their syntheses will produce mismatches which mean they are less likely to anneal and form a ligated product.

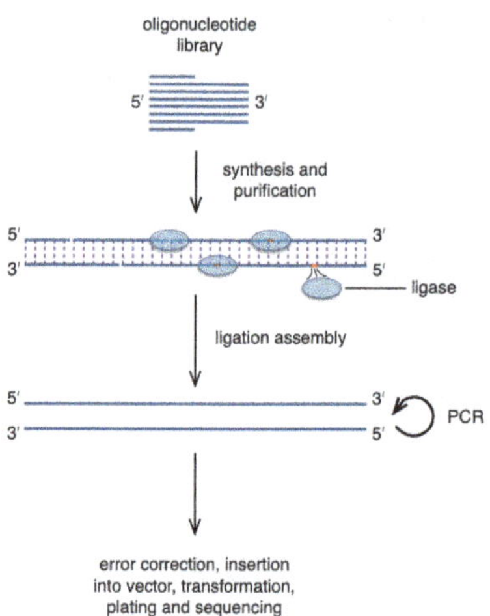

Gene synthesis by ligationOligos.

Solid phase synthesis generates the constituent fragments which are then purified ready for ligation. These oligos span the entire gene on both strands, missing only the phosphodiester bonds that link neighbouring oligos. Fragments are designed so two adjacent oligos are held together with a complementary template (the splint) then ligatedenzymatically.

Three major methods of ligation-based synthesis exist:

- Shotgun ligation – the gene is synthesised en masse by simply mixing the constituent oligos together and adding a ligase. The reaction can occur at room temperature though it usually requires extended reaction times. Use of thermo stable ligases and high purity oligos have now made this one-pot reaction more efficient.
- Ligase chain reaction (LCR) – LCR originally superseded shotgun ligation as a more efficient method with shorter reaction times and the ability to ligate longer genes (still limited to 2 kbp). This method follows a similar set of melting and annealing steps to PCR though differs by using a ligase instead of a polymerase to ligate templated oligos until the entire gene is assembled.
- Solid-support based ligation-mediated oligonucleotide assembly – The initial dsDNA oligo is immobilised on a solid support and iterative rounds of annealing and ligation steadily assemble the entire gene, which is cleaved from the support at the end of the synthesis.

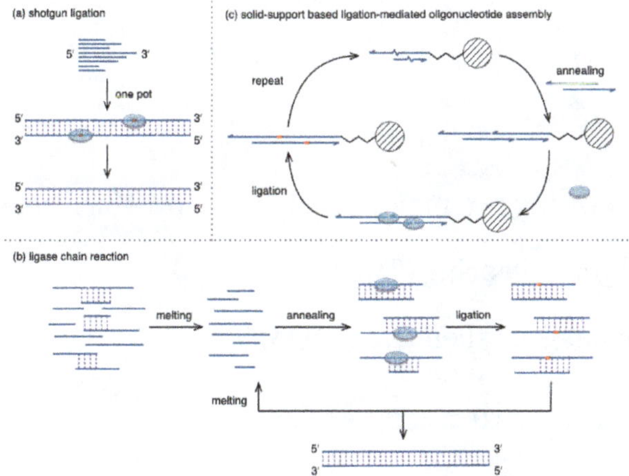

Three methods of ligation-based gene synthesis.

While the error rate of ligation based synthesis is lower, it may still be necessary to perform error correction to yield a legitimate number of viable genes. Furthermore, PCA is the dominant methodology used for artificial gene synthesis. Ligation-based techniques require polyacrylamide gel electrophoresis (PAGE) purification and 5' phosphorylation, making the whole method costly and labor intensive. PCA and associated error correction is preferred as these steps are eliminated.

Further methods of gene synthesis exist including assembly directly on to the plasmid relying on bacterial homologous recombination (Sequence- and Ligation- Independent Cloning – SLIC) or in vivo assembly of transformed oligonucleotides by the yeast Saccharomyces cerevisiae. PCA and ligation based methods remain by far the most common methods of gene synthesis.

Gene Silencing

The term gene silencing is commonly used to describe the "switching off" of a gene by a mechanism without genetic modification. The term gene silencing refers to an epigenetic phenomenon, the heritable inactivation of gene expression that does not involve any changes to the deoxyribonucleic acid (DNA) sequence. While this phenomenon has initially been studied in transgenic plants, its relevance in the regulation of endogenous plant genes has become increasingly apparent.

Below some cellular components are mentioned where gene silencing occurred:

- Chromatin and heterochromatin
- Dicer
- dsRNA
- Histones
- MicroRNA
- siRNA
- Transposons

Gene silencing has following two major subdivisions by which genes are regulated:

1. Transcriptional gene silencing (TGS);
2. Posttranscriptional gene silencing (PTGS).

Transcriptional Gene Silencing (TGS)

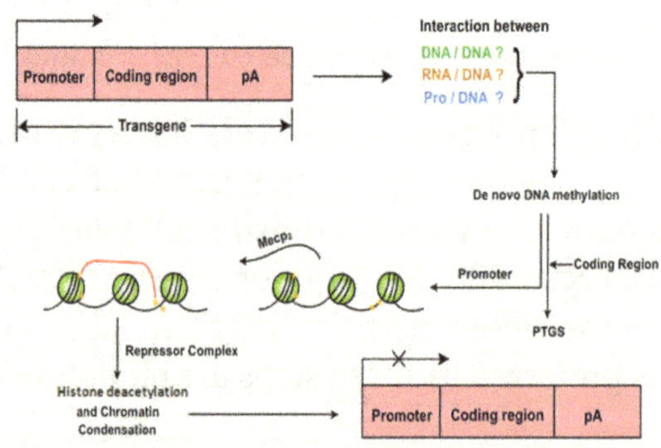

Schematic representation of DNA methylation – mediated transcriptional gene silencing (TGS).

Transcriptional gene silencing is the product of chromosomal histone modifications, creating an environment of heterochromatin, which is surrounded to a gene that makes it inaccessible to transcriptional machinery (RNA polymerase, transcription factors, etc.). TGS blocks primary transcription from nuclear DNA and is in most cases associated with DNA methylation and chromatin condensation in nearly all organisms that possess a DNA methylation system.

Post-transcriptional Gene Silencing (PTGS)

Post-transcriptional gene silencing is the product of transcribed mRNA of a specific gene being silenced. When mRNA was destructed, then translation to form an active gene product (in most cases, a protein) will be prevented. A general process of post-transcriptional gene silencing is by RNAi. PTGS involves a cytoplasmic, target sequence-specific RNA degradation process that is possibly activated by double-stranded RNA (dsRNA). This dsRNA is independent of ongoing translation.

TGS can be transmitted generation to generation by meiosis whereas PTGS is usually lost during meiosis. In PTGS, double stranded RNA is interred into a cell and gets chopped up by the enzyme known as dicer to form siRNA. siRNA then binds to the RNA-induced silencing complex (RISC) and is unwound. The antisense. RNA complex with RISC protein and binds to its corresponding mRNA, which is then cleaved by the enzyme slicer rendering it inactive.

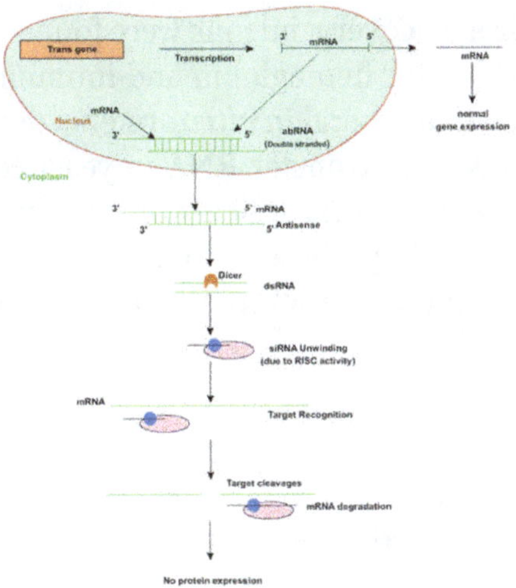

Schematic representation of post translational gene silencing (PTGS).

RNA Splicing

For most eukaryotic genes (and some prokaryotic ones), the initial RNA that is transcribed from a gene's DNA template must be processed before it becomes a mature messenger RNA (mRNA) that can direct the synthesis of protein. One of the steps in this processing, called RNA splicing, involves the removal or "splicing out" of certain sequences referred to as intervening sequences, or introns. The final mRNA thus consists of the remaining sequences, called exons, which are connected to one another through the splicing process. RNA splicing was initially discovered in the 1970s, overturning years of thought in the field of gene expression.

Early Studies in Bacteria

Gene regulation was first studied most thoroughly in relatively simple bacterial systems. Most bacterial RNA transcripts do not undergo splicing; these transcripts are said to be co-linear, with DNA directly encoding them. In other words, there is a one-to-one correspondence of bases between the gene and the mRNA transcribed from the gene (excepting 5′ and 3′ noncoding regions). However, in 1977, several groups of researchers who were working with adenoviruses that infect and replicate in mammalian cells obtained some surprising results. These scientists identified a series of RNA molecules that they termed "mosaics," each of which contained sequences from noncontiguous sites in the viral genome. These mosaics were found late in viral infection. Studies of early infection revealed long primary RNA transcripts that contained all of the sequences from the late RNAs, as well as what came to be called the intervening sequences (introns).

Subsequent to the adenoviral discovery, introns were found in many other viral and eukaryotic genes, including those for hemoglobin and immunoglobulin. Splicing of RNA transcripts was then observed in several *in vitro* systems derived from eukaryotic cells, including removal of introns from transfer RNA in yeast cell-free extracts. These observations solidified the hypothesis that splicing of large initial transcripts did, in fact, yield the mature mRNA. Other hypotheses proposed that the DNA template in some way looped or assumed a secondary structure that allowed transcription from noncontiguous regions.

Occurance of Splicing

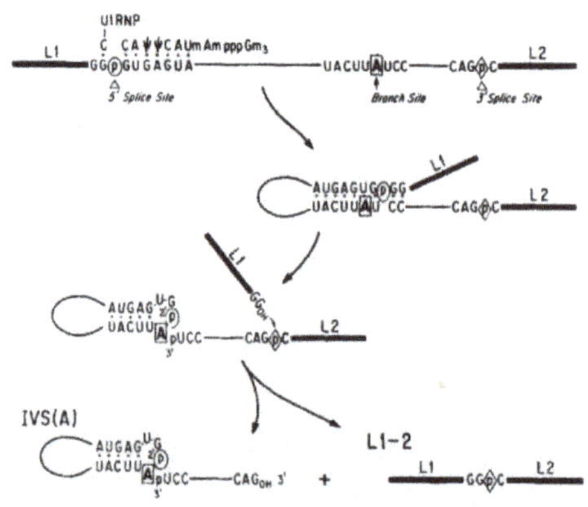

Pre-mRNA splicing

The biochemical mechanism by which splicing occurs has been studied in a number of systems and is now fairly well characterized. Introns are removed from primary transcripts by cleavage at conserved sequences called splice sites. These sites are found at the 5′ and 3′ ends of introns. Most commonly, the RNA sequence that is removed begins with the dinucleotide GU at its 5′ end, and ends with AG at its 3′ end. These consensus sequences are known to be critical, because changing one of the conserved nucleotides results in inhibition of splicing. Another important sequence occurs at what is called the branch point, located anywhere from 18 to 40 nucleotides upstream from the 3′ end of an intron. The branch point always contains an adenine, but it is otherwise loosely conserved. A typical sequence is YNYYRAY, where Y indicates a pyrimidine, N denotes any nucleotide, R denotes any purine, and A denotes adenine. Rarely, alternate splice site sequences are found that begin with the dinucleotide AU and end with AC; these are spliced through a similar mechanism.

Splicing occurs in several steps and is catalyzed by small nuclear ribonucleoproteins (snRNPs, commonly pronounced "snurps"). First, the pre-mRNA is cleaved at the 5′ end of the intron following the attachment of a snRNP called U1 to its complementary sequence within the intron. The cut end then attaches to the conserved branch point region downstream through pairing of guanine and adenine nucleotides from the 5′ end and the branch point, respectively, to form a looped structure known as a lariat. The bonding of the guanine and adenine bases takes place via a chemical reaction known as transesterification, in which a hydroxyl (OH) group on a carbon atom of the adenine "attacks" the bond of the guanine nucleotide at the splice site. The guanine residue is thus cleaved from the RNA strand and forms a new bond with the adenine.

Next, the snRNPs U2 and U4/U6 appear to contribute to positioning of the 5′ end and the branch point in proximity. With the participation of U5, the 3′ end of the intron is brought into proximity, cut, and joined to the 5′ end. This step occurs by transesterification; in this case, an OH group at the 3′ end of the exon attacks the phosphodiester bond at the 3′ splice site. The adjoining exons are covalently bound, and the resulting lariat is released with U2, U5, and U6 bound to it.

In addition to consensus sequences at their splice sites, eukaryotic genes with long introns also contain exonic splicing enhancers (ESEs). These sequences, which help position the splicing apparatus, are found in the exons of genes and bind proteins that help recruit splicing machinery to the correct site. Most splicing occurs between exons on a single RNA transcript, but occasionally trans-splicing occurs, in which exons on different pre-mRNAs are ligated together.

The splicing process occurs in cellular machines called spliceosomes, in which the snRNPs are found along with additional proteins. The primary variety of spliceosome is one of the most plentiful structures in the cell, and recently, a secondary type of spliceosome has been identified that processes a minor category of introns. These introns are referred to as U12-type introns because they depend upon the action of a snRNP called U12 (the common introns described above are called U2-type introns). The role of U12-type introns is not yet defined, but their persistence throughout evolution and conservation between homologous genes of widely divergent species suggests an important functional basis.

Self-Splicing and Alternative Splicing

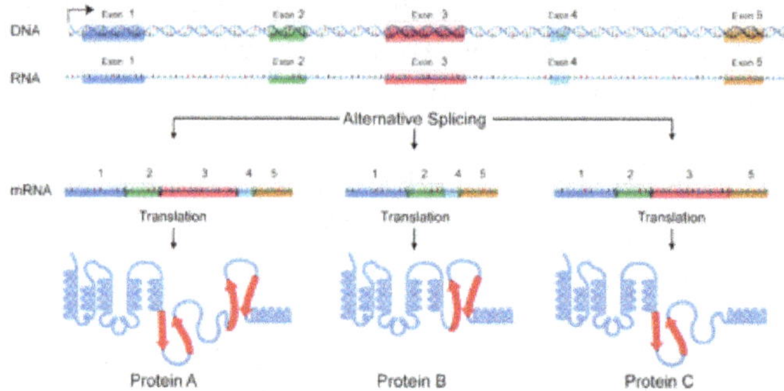

A schematic representation of alternative splicing

Some RNA molecules have the capacity to splice themselves; the initial discovery of this self-splicing ability in the protozoan *Tetrahymena thermophila* was recognized with the Nobel Prize in 1989. The self-splicing introns found in *T. thermophila* are now referred to as Group I introns; this class also includes other protozoan ribosomal RNA genes, some fungal mitochondrial genes, and some phage genes. Group I introns all fold into a complex secondary structure with nine loops and employ transesterification reactions as described above. On the other hand, Group II self-splicing introns are found in mitochondrial genes and are excised by a mechanism that bears similarities to pre-mRNA splicing, including the production of lariats. For this reason, it has been proposed that perhaps pre-mRNA introns and splicing mechanisms evolved from the Group II introns.

Early in the course of splicing research, yet another surprising discovery was made; specifically, researchers noticed that not only was pre-mRNA punctuated by introns that needed to be excised, but also that alternative patterns of splicing within a single pre-mRNA molecule could yield different functional mRNAs.

The first example of alternative splicing was defined in the adenovirus in 1977 and demonstrated that one pre-mRNA molecule could be spliced at different junctions to result in a variety of mature mRNA molecules, each containing different combinations of exons.

Shortly afterward, alternative splicing was found to occur in cellular genes as well, with the first example identified in the *IgM* gene, a member of the immunoglobulin superfamily. Another example of a gene with an impressive number of alternative splicing patterns is the *Dscam* gene from *Drosophila*, which is involved in guiding embryonic nerves to their targets during formation of the fly's nervous system. Examination of the *Dscam* sequence reveals such a large number of introns that differential splicing could, in theory, create a staggering 38,000 different mRNAs. This ability to create so many mRNAs may provide the diversity necessary for forming a complex structure such as the nervous system. In fact, the existence of multiple mRNA transcripts within single genes may account for the complexity of some organisms, such as humans, that have relatively few genes (approximately 20,000). For example, work from Wang *et al.* (2008) suggests that more than 90% of human genes are alternatively spliced.

Alternative Splicing

Alternative splicing of precursor mRNA is an essential mechanism to increase the complexity of gene expression, and it plays an important role in cellular differentiation and organism development. Regulation of alternative splicing is a complicated process in which numerous interacting components are at work, including cis-acting elements and trans-acting factors, and is further guided by the functional coupling between transcription splicing. Additional molecular features, such as chromatin structure, RNA structure and alternative transcription initiation or alternative transcription termination, collaborate with these basic components to generate the protein diversity due to alternative splicing. All these factors contributing to this one fundamental biological process add up to a mechanism that is critical to the proper functioning of cells. Any corruption of the process may lead to disruption of normal cellular function and the eventuality of disease. Cancer is one of those diseases, where alternative splicing may be the basis for the identification of novel diagnostic and prognostic biomarkers, as well as new strategies for therapy. Thus, an in-depth understanding of alternative splicing regulation has the potential not only to elucidate fundamental biological principles, but to provide solutions for various diseases.

Molecular Mechanisms of Alternative Spicing

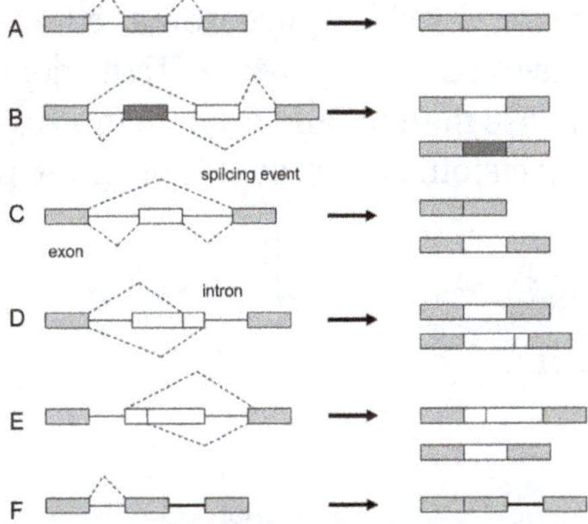

Five main types of alternative splicing events are depicted. (A) Constitutive splicing; (B) mutually exclusive exons; (C) cassette alternative exon; (D) alternative 3′ splice site; (E) alternative 5′ splice site; and (F) intron retention.

Systematic analyses of ESTs and microarray data have so far revealed seven main types of alternative splicing. The most prevalent pattern (~30%) is the cassette-type alternative exon (exon skipping) in vertebrates and invertebrates, while in lower metazoans, it is intron retention. Intron retention in human transcripts is positioned primarily in the un-translated regions (UTRs) and has been associated with weaker splice sites, short intron length and the regulation of cis-regulatory elements.

Alternative selection of 5′ or 3′ splice sites within exon sequences (~25%) may lead to subtle changes in the coding sequence, and an additional layer of complexity arises with mutually exclusive alternative exons. One example of a transcript that undergoes alternative splicing, which generates variation in the protein. Differences in the splicing machinery in different cell types and unique cis-acting elements in the *FGF-R2* pre-mRNA lead to altered tissue specific choices that create either *FGF-R2IIIb* or *FGF-R2IIIc* mature transcripts.

The protein expression is further regulated by alternative polyadenylation of mRNA, which influences the coding potential or the 3′UTR length by modifying the binding availability of microRNA or RNA. Of note, it has been demonstrated that each type of alternative splicing can operate in a stochastic manner, and different splice-site identification and processing mechanisms do not necessarily occur at the same frequencies among all biological kingdoms.

The mechanisms outlined above are just one indication of the complexity, as numerous molecules are involved in alternative splicing in a coordinated manner. Even the basic nucleotide components and the essential molecules that recognize them can introduce diversity in the synthesis of mature transcripts.

Two major steps constitute the basic process of splicing: Assembly of the spliceosome followed by the actual splicing of pre-mRNA. The spliceosome is mainly composed of U1, U2 small nuclear ribonucleic proteins (snRNPs) and the U4/U6.U5 tri-snRNP, and configure in identify a core set of splicing signals: The 5′ splice site, the branch point sequence and the 3′ splice site. Specific spliceosomal complexes (E, A, B and others) and eight evolutionarily conserved DExD/H-type RNA- dependent ATPases/helicases assemble in a proposed stepwise manner and execute multiple splicing steps that result in exon ligation and intron excision. Numerous steps in the pathway are reversible.

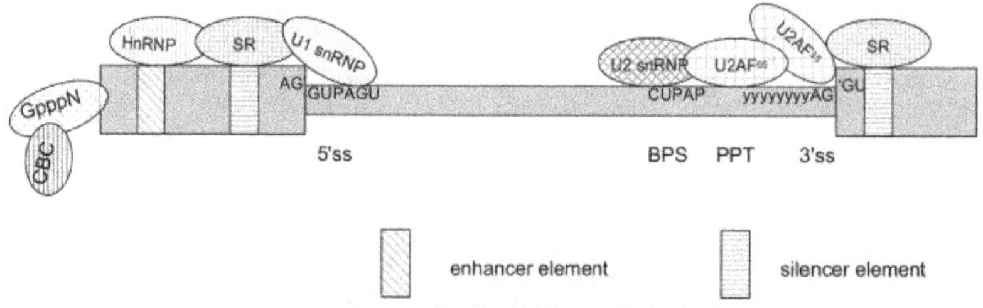

Schematic representation of the sequence elements and proteins at 5′ and 3′ exon-intron boundaries in an RNA transcript.

The exons that end up in the mature mRNA during the process of alternative splicing is entirely defined by the interaction between cis-acting elements and trans-acting factors. Cis-acting elements include exonic splicing enhancers (ESEs) and intronic splicing enhancers (ISE) that are bound by positive trans-acting factors, such as SR proteins (serine/arginine-rich family of nuclear phosphoproteins), whereas exonic splicing silencers (ESSs) and intronic splicing silencers are bound by negative acting factors, such as heterogeneous nuclear ribonucleoproteins (hnRNPs). The collaboration between these elements results in the promotion or inhibition of splicesome assembly of the weak splice sites, respectively. In general, the cis-acting elements function additively. The enhancing elements tend to play dominant roles in constitutive splicing, while the silencers are relatively more important in the control of alternative splicing. Enhancer activity has been shown to be abolished by a stable stem-loop structure as short as 7 base pairs in an RNA transcript owing to the mechanisms of physical competition, long-range RNA pairing, a structure splice code and co-transcription splicing. Furthermore, the specificity of cis-acting enhancer elements for introns or exons has been investigated. In these experiments, an ESE was found to act as an ISE depending on its location in an exon or intron.

HnRNPs are highly conserved from nematodes to mammals and have several critical roles in pre-mRNA maturation. Their function is to bind to the ESS to the exclusion of SR proteins. A looping out pre-mRNA leads to exonic sequestration from the rest of pre-mRNA transcript. HnRNPs A/B are a family of RNA-binding proteins, its diversification roles in the modulation of alternative splicing have evolved based on differing affinities for their cognate nucleic acids. HnRNP H and F serve to alter the proteolipid protein (PLP/DM20) ratio via the variation in the recruitment of U1 snRNP. Similarly, the antagonistic role of hnRNP M to the splicing factor Nova-1 generates alternatively spliced dopamine receptor pre-mRNAs, which create isoforms associated with diverse key physical functions, such as control, reward, learning and memory. In addition, hnRNP L and phosphorylation of ser513 have been recently shown to be involved in the regulation of alternative splicing through dynamic membrane depolarization and Ca^{2+}/ calmodulin-dependent protein kinase IV activation.

In addition to the coupling of SR proteins to enhancer elements, SR proteins interact with U1 snRNP and the 35 kDa subunit of the heterodimeric factor, U2AF. The second subunit of U2AF, $U2AF^{65}$, binds SF1 and the pyrimidine tract simultaneously, on the basis of the arginine/serine (RS)-rich domain, which results in recognition and stability of the branch point, as well as polypyrimidine tract sequences. Approximately 10–12 serines in the N-terminal region of the RS domain are rapidly phosphorylated by the binding of SR-specific protein kinase to serine/arginine-rich splicing factor 1 with an unusually high affinity. This continuous phosphorylation/dephosphorylation cycle of SR proteins facilitates the shuttling of SR proteins between the nucleus and the cytoplasm, and is critically required for the regulation of alternative splicing by growth signals transduced to the nucleus. SR proteins have also been proposed to participate in post-splicing activities, such as mRNA nuclear export, nonsense-mediated decay (NMD) and mRNA translation.

In general, positive or negative splice-site recognition is regulated through various mechanisms, such as the local concentration or activity of splicing regulatory factors, under diverse physiological or pathological conditions.

Coupling of Alternative Splicing to Transcription

A large number of components associated with the physical interaction between splicing and transcription have been purified, with particular attention on the carboxyl terminal domain (CTD) of the large subunit of RNAPII. The CTD consists of 52 tandem repeats of the heptapeptide YSPTSPS in mammals (26 tandem repeats in yeast), which act as a special platform to recruit different factors to the nascent transcripts via dynamic phosphorylation of serine residues. Kinases that phosphorylate specific CTD serine residues have been identified and are components of the protein apparatus driving the specific function. For example, ser5 phosphorylation is associated with transcription initiation through cyclin-dependent kinase 7 (CDK7) of the general transcription factor IIH (TFIIH), whereas ser2 phosphorylation is preferentially linked with CTD activity at the 3′-end of genes through CDK9 of the positive transcription elongation factor. In addition, phosphorylation of ser7 has been found to facilitate elongation and splicing. Thus, phosphorylation is a mechanism that clearly demonstrates that functional coupling exists between transcription and alternative splicing.

CTD participates in gene expression-related functions ranging from 5′ capping, splicing, poly-adenylation and chromatin remodeling, Of note, mutation and deletion analysis of CTD has revealed multiple defects in mRNA processing therefore, CTD and additional components of the two machineries have emerged as a central element in governing the interactions between transcription and splicing. Taken together, functional coupling appears to maintain an important role in alternative splicing in driving determinative physiological changes, and fine-tune gene expression in mathematical modeling approaches.

Two models have been suggested to explain the co-transcription process of how transcription coupled repair influences alternative splicing. The mechanism of the recruitment model may mainly depend on specific features of CTD (as mentioned above), whereas the kinetic model is based on the different elongation rates of Pol II, which in turn determine the timing of the presentation of splices sites.

Fundamentally, the aforementioned mechanism influences patterns of alternative splicing via the variations in Pol II elongation and recruitment of splicing factors by specific histone marks. Thus, alternative splicing is highly influenced not only by transcription, but also by the chromatin structure, which underscores chromatin as another layer in the regulation of alternative splicing. The resultant mature mRNA is thus a reflection of numerous DNA modifications, such as patterns of histone methylation at exons, modulation of histone modifications and increased DNA methylation at exons, Conversely, a previous study indicated that splicing may mediate chromatin remodeling via deposition of histone marks on DNA or numerous associations between splicing factors and elongation proteins.

Adding additional complexity to the regulation network is alternative transcription initiation (ATI) and alternative transcription termination (ATT) sites. ATI and ATT significantly contribute to the diversity of the human and mouse transcriptomes to a degree that may exceed alternative splicing, when considering the number of possibilities available through alternative nucleotides, isoforms and introns. In contrast to the prevalence of alternative splicing that occurs within coding sequences (CDSs), the dominant class of alternative events, which includes ATI and ATT, occur in UTRs. This discovery reflects the preferential regulation of large distinct groups of genes with different mechanisms, such as strong coupling with alternative splicing in 5′ and 3′UTRs.

Despite the strong correlation between alternative splicing and transcription, alternative transcription mainly results in variations of the transcript number or the 5′/3′ terminal protein variants due to differential transcriptional start or terminal sites. By contrast, alternative splicing associated alterations mostly lie within the protein sequence, potentially affecting almost all areas of protein function.

Trans-splicing

Trans-splicing is a common phenomenon in trypanosomes, nematodes, *Drosophila* and even humans, and refers to the novel and unusual splicing of exons from independent pre-mRNAs. The phenomenon has been explored as a therapeutic option for a variety of genetic diseases, particularly in the treatment of cancer. The carcinoembryonic antigen (CEA), for example, is associated with a variety of neoplastic processes and was exploited as a target for trans-splicing. A CEA RNA-targeting trans-splicing ribozyme was designed to perform RNA replacement through a trans-splicing reaction specifically in CEA expressing cells. The activity of the ribozyme simultaneously reduced CEA expression and introduced the thymidine kinase gene, which rendered the cells sensitive to ganciclovir treatment. RNA trans-splicing has also been utilized for the potential treatment of neurodegenerative diseases through a novel technology, splic osome mediated trans-splicing (SMaRT). SMaRT was successfully used *in vivo* to re-engineer tau mRNA transcripts to include E10, and therefore, offers the opportunity potential to correct tau missplicing and treat the underlying disease.

Epitranscriptome

An epitranscriptome is a set of functionally appropriate RNA modifications. There are nearly one hundred known modifications of RNA; the most common modification in internal mRNA is N6-methyladenosine (m^6A). Transfer RNAs (tRNAs), mRNAs, and ribosomal RNAs (rRNAs) are the three types of RNAs that are post-transcriptionally modified.

DNA and RNA helix.

In all living organisms, translation is a fundamental process. The arrangement and operation of translation process is very expensive and it consumes 40% of cellular energy, which has led to the need for strict regulation of protein production in many aspects.

The regulation of translation is generally related to the need of regulated and non-regulated proteins. Nucleotide modifications are also important for the translation process. RNA base modifications are detected by appropriate sequencing methods.

At a genome level, the study of m^6A is possible with the help of recent advances in sequencing technology and the design of m^6A antibodies. The sequencing technologies are listed below.

m^6A-sequencing

The m6A is the most available base in eukaryotic mRNA. The m6A protocol can be carried out on mRNA. The first step is to acquire a better quality of RNA sample. Polyadenylated RNA is enhanced by oligo-dT selection while the m6A-seq protocol is executed through RNA. Polyadenylated RNA is divided into small fragments with a length of 100 nt each.

Polyadenylated RNA fragmentation is performed chemically to achieve efficient fragmentation. These small fragments are used to reduce the resolution to 200 nt. Remapping of smaller fragments into the genome would become tedious because of further fragmentation.

Post-fragmentation, the sample is put through immunoprecipitation. This small amount of fragmented mRNA is retained aside to perform input control process. In the following data analysis, small fragments are retained at basic level status. The rest of the sample is immunoprecipitated by incubating it with a particular m6A antibody that pulls fragments containing multiple m6A residues.

Recombinant protein agarose bead, which can bind with the antibodies, is used to incubate the RNA-antibody mixture in the subsequent process. Finally, the unbound fragments are eradicated, bound fragments are removed by addition of m6A to the mixture, and the bead-antibody-RNA complex deposited in solid form.

During immune precipitation, the RNA fragments and beads are combined together in the absence of antibodies. Bioanalyzer analysis and concentration measurements are used to estimate the quality of the material throughout the remaining protocol. Good result can be obtained when better quality starting material is used.

MeRIP-sequencing

There exist some differences between m6A sequencing and MeRIP-sequencing strategies. The main difference is the category of bead: Magnetic Dynabeads are used in MeRIP sequencing strategy, which is first paired with antibodies and then combined with ribo-minus treated RNA. Here, immune precipitation process is performed twice.

Buffer composition involved in fragmentation also differs. Solvent extraction elution is used instead of competition elution. However, the fundamental concept involved in these two methods is similar to each other.

PA-m6A-sequencing

The PA-m6A sequencing name was coined from m6A sequencing strategy in combination with a PA sequencing strategy. In this method, PA-m6A has the ability of undergoing high-resolution mapping with a mammalian transcriptome m6A. It can be observed that the m6A sequence modification sites are more accurately defined when compared to conventional sequencing methods. The resolution has been reduced from 200 to 23 nt.

PA-m6A sequence is versatile when compared to the single nucleotide approaches. Photoactivatable ribonucleoside-enhanced crosslinking and immunoprecipitation (PARCLIP) strategy is the base for PA-m6A strategy. In PARCLIP strategy, photoactivatable ribonucleosides are integrated into an mRNA and then they are subjected to immuneoprecipitation process. It is similar to m6A protocol, but immunoprecipitation is performed with full length RNA and then this RNA is fragmented post-crosslinking through UV irradiation.

Single Molecule Sequencing

Single molecule sequencing has the ability to provide the base level resolution of m6A sites without the conclusion drawn on the basis of motif. The Single Molecule Real-Time (SMRT) technology is the most common platform for this method of sequencing.

This method was developed by Pacific Biosciences and Oxford Nano pore Technologies. Continuous monitoring of the process, in which DNA polymerase is integrated with fluorescent nucleotides, helps in detecting both genetic and epigenetic information at the same time.

Single molecule sequencing offers some advantages as it can be used to detect RNA modification that includes m6A sites. Single nucleotide resolution is to detect the m6A sites in RNA, which is achieved by using DNA polymerase rather than reverse transcriptase.

It can be used as an enzyme within a ZMW and in allowing direct complementary DNA (cDNA) synthesis in real time. Reverse transcription appears at standard speeds. At the instant of base incorporation, designed m6A sites exhibit important growth of inter pulse duration (IPD).

Single Nucleotide Methods

The above techniques are not useful in the case of determining nucleotide level modification, due to a lower determination capability. Site-specific cleavage and radioactive-labeling followed by ligation-assisted extraction and thin-layer chromatography method can be used to overcome this disadvantage. Therefore, the m6A modification is non-stoichiometric.

References

- Gene-Expression-An-Overview: news-medical.net, Retrieved 18 April 2018
- Genetic-code: biologydictionary.net, Retrieved 28 June 2018
- Antitermination-control-of-gene-expression-molecular-biology: what-when-how.com, Retrieved 19 May 2018
- Gene-synthesis, Artificial-gene-synthesis-63: atdbio.com, Retrieved 26 June 2018
- Epitranscriptome-Sequencing-Technologies, life-sciences: news-medical.net, Retrieved 18 July 2018

Chapter 7

Molecular Evolution

Any change in the sequence composition of cellular molecules like DNA, RNA and proteins over multiple generations is referred to as molecular evolution. Some of the major topics of study in this domain include the genetic basis of speciation, origin of new genes, natural selection vs. neutral evolution, genetic nature of complex traits, etc. This chapter delves into the fundamentals of mutation, gene duplication and molecular evolution.

Changes in the base pair sequences in DNA or RNA molecules and changes in amino acid sequences and their molecular configuration in different proteins, from generation to generation are known as molecular evolution.

It is possible to measure differences between these molecules obtained from different organisms (such as humans, apes, monkeys, prosimians etc.) on a unit scale of amino acids or nucleotides and demonstrate their relationships. As the molecular sequences are heritable, their variations produce molecular records that have been transferred from generation to generation during evolution.

A triplet made of three pairs of nucleotides is called a codon. A codon will change if one of the three bases changes and it may or may not end up in a change in the amino acid synthesized by it. Majority of these changes are small and inconsequential but accumulate over long periods to bring about large alterations in the gene frequencies in populations. Two kinds of such changes are possible:

Silent site substitution: These are such changes in DNA sequences which do not result in any change in amino acid synthesis and hence composition of proteins is not changed. They are usually changes in the last base pair of the codon. For example in mRNA strand GCA codes for alanine and if adenine is replaced by guanine, the resulting GCG will still code for the same amino acid alanine. Silent site substitutions do not bring about any phenotypic changes.

Replacement substitution: They are changes in the bases of codons that result in synthesis of new amino acids and are capable of altering the structure of proteins that are controlled by them and thus changing the phenotype.

Silent site substitutions have much higher rate of change as compared to the replacement substitutions, since the former do not produce changes that can be exposed to natural selection but the latter do. For the same reason genes which are less vital to the cell can undergo rapid changes by replacement substitution without showing harmful effects. Pseudo genes, which are duplicated sequences of bases and do not code for proteins and hence are not exposed to natural selection, are known to undergo higher rate of evolutionary changes.

Sequencing amino acids: Comparing amino acid sequences in a protein in different species by using biochemical techniques is one of the most popular methods to determine phylogeny. For example, in haemoglobin two pairs of alpha and beta sequences of polypeptide chains form a tetramer that can be distinguished by different amino acid sequences in different species. In vertebrates different types of globin chains appeared during evolution and in each species they followed their own evolutionary path by changes in the amino acid sequences. They are all variations of a single globin ancestor that is controlled by similar globin genes which are believed to have originated by gene duplication of the original type.

Neutral Theory of Molecular Evolution

Motoo Kimura proposed that a vast majority of base substitutions that are preserved in a population are neutral with regards to natural selection. Positive substitutions are so rare that they are inconsequential in molecular evolution, while negative changes are quickly eliminated by natural selection. Natural selection seems to favor neutral changes which determine the overall rate of sequential evolution. For instance, pseudo genes have the highest substitution rate among the genes but the changes are completely neutral with regard to selection.

The theory was tested by J. McDonald and M. Kreitman by comparing base sequences of alcohol dehydrogenase gene of *Drosophila melanogaster, D. simulans* and *D. yakuba.*

Kimura's theory not only contradicts classical Darwinism but also does not explain fixation of various types of alleles in different sizes of population. The theory holds that the rate of fixation of neutral mutations does not depend on population size but the genes are fixed or eliminated by genetic drift.

The neutral theory provides theoretical framework for testing and predicting molecular evolution in the absence of positive selection.

The Molecular Clock

E. Zuckerkandl and Linus Pauling (1962) found that the rate of divergence in aminoacid sequences of haemoglobin and cytochrome **c** in different species-pairs in mammals always remained constant. In other words, changes in the base sequences of DNA and the resultant amino acid substitution accumulates in a population in a clock-like regularity over a period of time and hence could be used to date branching evolution or cladogenetic events. This was called *molecular clock,* which before putting to use needs to be calibrated by matching observed genetic divergence of the living population with the absolute time of divergence as revealed by the fossil records. Differentiation in haemolymph proteins of Hawaiian drosophilids gave an idea of splitting of phyletic lines and colonization of Hawaiian Islands by these flies about 40 million years ago from North America.

Sarich & Wilson used this method to find out the divergence of hominids from apes by calibrating the amount of molecular differentiation achieved between the two groups in relation to time by taking the example of divergence of Old World and New World monkeys. The measurement of divergence in albumin gave the time of split of hominids and apes at 5 million years from present, which is supported by other evidences.

The following three types of changes are considered for molecular clock:

- Those base sequences in which substitution has taken place in the third position of codon, since they are expected to be neutral.
- Changes in pseudo genes are not exposed to natural selection and hence are likely to give better results.
- When natural selection is very strong, only neutral substitutions are likely to be fixed in a population and should only be considered for molecular clock.

Changes in Mitochondrial DNA are linear and constant like clock ticking and hence are commonly used in molecular clock. In mtDNA of mammals the divergence is about 2% per million years. In sea urchins this rate of divergence is estimated to be at 1.8-2.2%, which is remarkably similar to mammals. In sharks that have reliable molecular clock owing to well-documented fossil records, the rate of change is estimated to be 8 times slower than in mammals. Therefore, there is no universal molecular clock even for mtDNA of animals.

Drawbacks

In spite of being useful there are serious drawbacks in the use of molecular clock. Substitution rates are variable among genes in different species due to differences in generation time and rate of mutation. The same molecule may evolve at different rates in different evolutionary lines. Also, regulatory genes, introns, transposons and gene families may demonstrate considerable deviations in their rates of divergence. Therefore, the clock might work in a particular lineage but not work in others or might work at different rates in different lineages.

Gene Duplication

Gene duplication refers to the duplication of a segment of DNA that contains one or more genes. Gene duplication is the primary source of new genes in evolution, and duplicate genes form gene families that are abundantly found in almost all genomes. For these reasons, gene duplication has been a main focus of molecular evolutionary study for decades. Recent years have also seen examples of harmful gene duplication that causes diseases, which bring gene duplication to the attention of human and medical geneticists.

Gene duplications are considered to be major genetic basis for producing novel genetic variations. There are three types of gene duplications: Whole Genome (WGD), segmental and small scale duplications by tandem duplications. Example of gene duplications of gene duplications events are frequently found in vertebrate serpins, which are classified into six groups V1-V6. Several tandem duplications on the same locus lead into several paralogs for groups V1 on the human chromosomes and V2 on the human chromosome. In contrast, there are several serpins which are localized on a single gene in the chromosomal fragments like angiotensinogen [chromosomes 1], heparin cofactor II. C1 inhibitor and antithrom bin III respectively. Similar trends are also followed by serpins from invertebrates like urochord dates.

About five decades ago, Susumu Ohno proposed his famous hypothesis that early vertebrates have undergone two rounds of WGD events and which is known as Ohno's hypothesis or 2R-hypothesis. These events cause massive gene duplications that are the hallmarks of gene and functional innovation. There were not one who believed on this hypothesis in the beginning years; later same theory become the cornerstone of gene duplications and fates, special in the post-genomic era. It is now clear that second WGD has also occurred in fishes and this often called as Fish- Specific Genome Duplication (FSGD) or 3R-hypothesis and it is best exemplified in the Hox clusters duplications.

On the evolutionary scale, gene duplication events lead into several schemes.

This includes following schemes:

a. Nonfunctionalization is random loss of function in one of the two gene copies by pseudogenization. Large fraction of duplicated genes are pseudogenized over 100 MY in the rainbow trout.

b. Neofunctionalization is when one gene copy may retain the original function while the other acquires a novel, evolutionarily advantageous/adaptive function.

c. Subfunctionalization is after duplication, mutations may occur in both genes that specialize to perform complementary functions.

There are several puzzles about how duplicate genes are retained during evolution. Several models have been proposed over in last four decades. Classical duplication-degeneration-complementation / sub functionalization models do not invoke positive selection; however, this can impose higher rates for retaining duplicate genes in small populations only. Rodents have higher retentions rates duplicate genes and only few losses in comparison to humans which corroborates that positive selections are more instrumental players than previously assumed.

Suppose a condition where two redundant gene copies were retained in the genome without significant functional divergence, this can covenant increase genetic robustness against harmful mutations in the concerned species. Within multi gene families descended from a common ancestor, these genes possess similarities at the DNA level, which implies for similar functions. The tandemly duplicated genes which generate several paralogous on the same chromosomal fragments, exhibit the case of concerted evolution. Within this concept, all genes in a given group evolve coordinately by homologous recombination, which vanguard into gene conversion. These paralogs share higher sequence identities like anti-trypsin-like gene cluster in the human chromosome For large fraction of multi gene families, the evolutionary model of birth-and-death (aka gain-and-loss) is largely supporting model, which propound that protein sequence similarities within family members is pronounced by strong purifying selection and evolvements of individual genes are primarily occur only by synonymous substitutions.

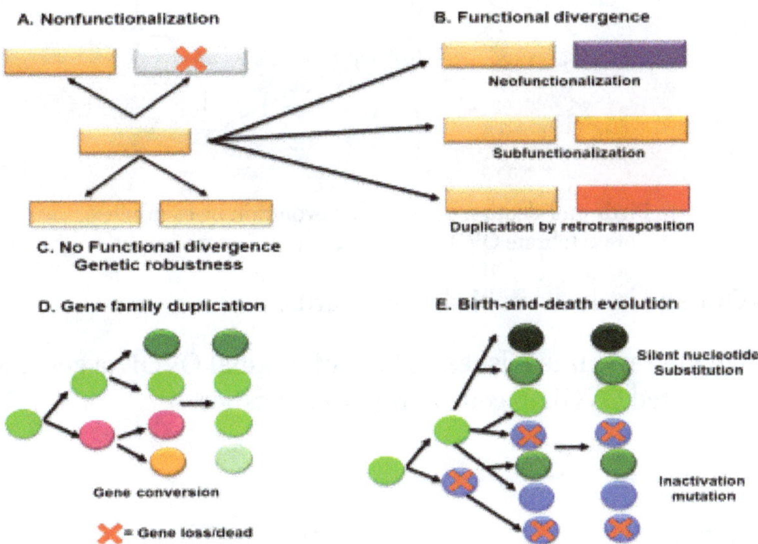

Fate of duplicated single genes (A-C) and duplicated gene families (D-E). Modified from Conrad and Antonarakis.

Vertebrate OVO-like transcription factors family depicts a good example of birth-and-death model of gene evolution. This family has lineage specific birth and dead of genes as reported earlier. Birds and lizards have loss of OVOL1 and its entire locus, while ray-finned fishes have loss of entire OVOL2 locus. Fishes have duplication of OVOL3, generating original OVOL3a and duplicated OVOL3b. Subsequently fishes lost OVOL3a and OVOL3b has compensated for this loss.

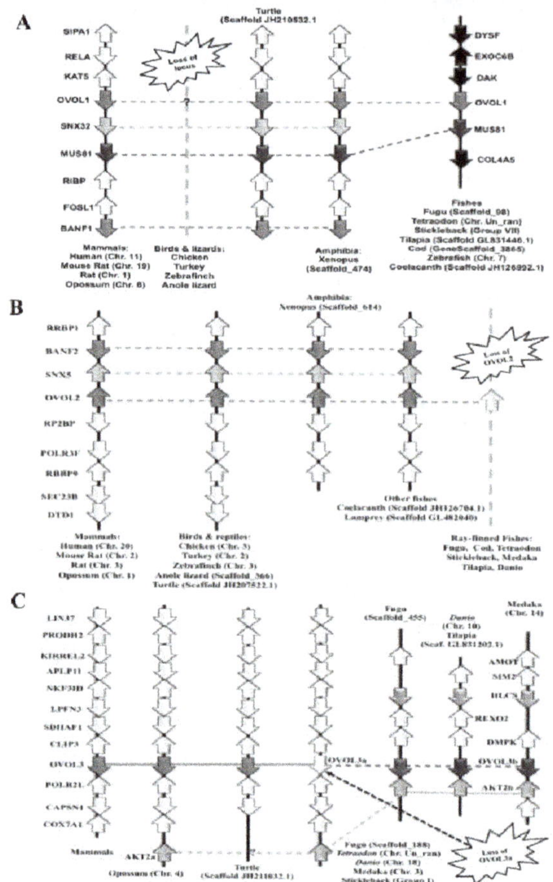

Birth-and-death model of gene evolution by example of vertebrate OVO-like transcription factors family

A) Loss of OVOL1 and its locus in birds and lizards.

B) Loss of OVOL2 in ray-finned fishes C) Loss of original OVOL3a gene, which is compensated by duplicated OVOL3b gene in ray-finned fishes.

Mutation

Mutations are alterations to a DNA sequence. If one thinks of the information in DNA as a series of sentences, mutations are errors in spelling the words that make up those sentences. Sometimes mutations are inconsequential, like a misspelled word whose meaning is still quite clear. At other times mutations have stronger ramifications, like a sentence whose meaning is completely changed.

Mutations are changes that occur in the nucleotide sequence of DNA. They can occur spontaneously when DNA is being replicated during cell division, but also can be induced by environmental factors, such as chemicals or ionizing radiation. Replication errors in human cells occur for every 100,000 nucleotides, which in turn amounts to about 120,000 errors each time one cell divides. However the good news is, in most cases, cells have the capacity to repair such errors. Or, the body destroys cells that cannot be repaired, thereby preventing a population of aberrant cells from expanding.

Types of Mutations

Broadly, mutations fall into two categories — somatic mutations and germ line mutations — Somatic mutations occur in their namesake somatic cells, which refers to the various cells of one's body that are not involved in reproduction; skin cells for example. If the replication of a cell with a somatic mutation is not stopped, then the population of aberrant cells will expand. However, somatic mutations cannot be passed on to an organism's offspring.

On the other hand, germ line mutations occur in the germ cells or the reproductive cells of multicellular organisms; sperm or egg cells for example. Such mutations can be passed on to an organism's offspring. Moreover, such mutations will carry over to pretty much every cell of an offspring's body.

However, based on *how* a DNA sequence is changed, many different types of mutations can occur. For instance, sometimes an error in DNA replication can switch out a single nucleotide and replace it with another, thereby changing the nucleotide sequence of only one codon. This type of error, also known as a base substitution can lead to the following mutations:

- Missense mutation: In this type of mutation the altered codon now corresponds to a different amino acid. As a result an incorrect amino acid is inserted into the protein being synthesized.
- Nonsense mutation: In this type of mutation, instead of tagging an amino acid, the altered codon signals for transcription to stop. Thus a shorter mRNA strand is produced and the resulting protein is truncated or nonfunctional.

Silent Mutations

A silent mutation is a change in the sequence of nucleotide bases which constitutes DNA, without a subsequent change in the amino acid or the function of the overall protein. Sometimes a single amino acid will change, but if it has the same properties as the amino acid it replaced, little to no change will happen. A silent mutation can be caused many ways, but the key point is that it does not change the function of the amino acid or subsequent proteins. A silent mutation is just that: it does nothing significant, not making a sound in the orchestra of the cell.

Silent Mutation Examples:

The Redundant Genome

The DNA is read in units of three nucleotides, called *codons*. Each codon specifies a certain amino acid, with a few reserved as stop and start signals. Sometimes, different codons specify the same amino acid. This redundancy allows flexibility in the genetic code. This means that a silent mutation usually goes completely unnoticed. You can see a typical silent mutation in the graph below:

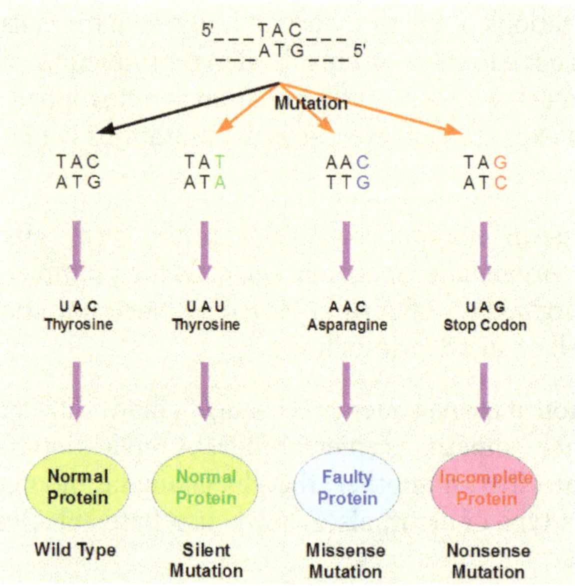

Here, a silent mutation is compared with both a *nonsense* mutation and a *missense* mutation. The silent mutation, which is an actual change at the DNA level from a *thymine* to a *cytosine*. This mutation could have been caused by a mistake in DNA replication, or from some sort of repair that happen after the DNA was damaged. Regardless both of these three nucleotide codons tell the *ribosome* and machinery within to attach a lysine amino acid.

In this case, the entire structure of the protein will remain the same regardless of the silent mutation. With the exact same amino acid structure, the protein will function no differently unless it is placed in a different environment. A silent mutation can also happen at the protein level, with no functional effect on the protein.

Amino Acid Groups

The 4 nucleotides, in groups of three codons, can call for all 21 amino acids. Seen below, the amino acids are grouped by their structure and side chains. These features directly impact how they interact with other amino acids, and what effects they have on molecules in the environment.

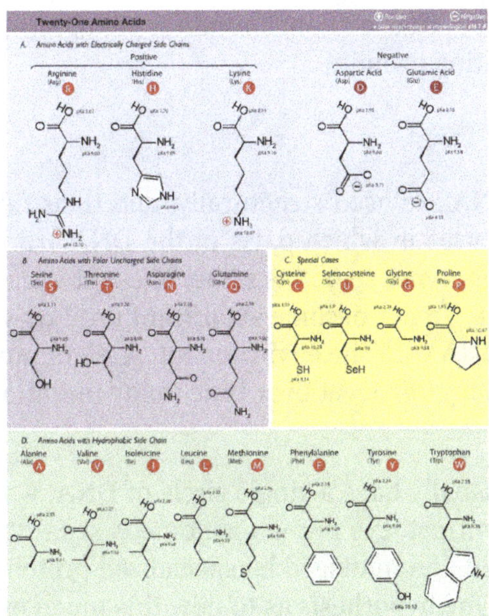

A silent mutation, which could easily include more than one nucleotide, could easily change an entire amino acid, or even series of amino acids. If a serine changed into a threonine, the effect might be minimal. The two amino acids are in the same category and are very similar shapes. This means that they will have a similar chemical reaction on the molecules around them. This will influence the shape and effect of the total protein. If the effect is negligible, the change is considered a silent mutation.

Place within Protein Structure

Several amino acids can be key to the entire structure or functionality of a protein. Many proteins have an *active site*, to which other molecules must bind. This site is constructed from a specific sequence of amino acids. When folded just right certain amino acids and their side chains will have the exclusive ability to interact with another molecule. If these amino acids are changed by a mutation, the functionality of bonding may be seriously impaired. This can change the function or utility of a protein.

Other proteins, on the inside of the molecule, have complex structures which must be present to preform specific functions. Many proteins undergo a *conformational change*, which is a change in shape. This is driven by electrical stimulation or the binding of a molecule like a *coenzyme* or a *substrate* to the protein. The conformational change, literally changing the shape of the protein, can press molecules together or tear them apart. The energy supplied is dependent upon the internal structure and specific bonds within the protein. Sometimes, a single amino acid can be a key piece of this. If this amino acid is changed for a non-functional one, the mutation is not a silent mutation. The change on the inside of the protein can also affect the functionality of the protein. Regardless of what a silent mutation changes, it should not change the functionality of the resulting protein.

Within Non-coding DNA

Many portions of the DNA are used structurally, and their full purpose is not understood. There are many cases in which parts of the DNA are vastly different between individuals, yet their phenotypes seem the same. These changes, especially small structural changes in the DNA, do not become significant until they begin to change the interaction of the coding DNA with the environment. A silent mutation could easily happen in these areas without notice, yet over time many mutations may begin to change a population.

Bacteria, interestingly, usually have a single circle of DNA, which carries all the information they need. By contrast, the human genome is separated on multiple chromosomes, which are bundled and managed by specialized proteins so they can be wound up during cell division. One hypothesis as to how this much more complex DNA came about was that certain silent mutations began forming structures of DNA. In a more compact genome, more information can be stored, which may have led to the complexity of life from single celled organisms to more complex forms. The folding and protection of various parts of DNA is part of normal *cell differentiation* in eukaryotes. Supposedly, these could have arisen through silent mutation until they became useful and were selected for.

Point Mutation

A point mutation is a type of mutation in DNA or RNA, the cell's genetic material, in which one single nucleotide base is added, deleted or changed. DNA and RNA are made up of many nucleotides. There are five different molecules that can make up nitrogenous bases on nucleotides: cytosine, guanine, adenine, thymine (in DNA) and uracil (in RNA), abbreviated C, G, A, T, and U. The specific sequence of nucleotides encodes all the information for carrying out all cell processes. In general, a mutation is when a gene is altered through a change in DNA structure; this may refer even to entire sections of chromosomes. A point mutation is specifically when only one nucleotide base is changed in some way, although multiple point mutations can occur in one strand of DNA or RNA.

From DNA to Protein

DNA and RNA have a double helix structure. Phosphate groups and 5-carbon sugars make up the backbone, while the middle of the double helix is formed by pairs of nitrogenous bases. Each type of nitrogenous base pairs with another specific base. Cytosine pairs with guanine, while adenine pairs with thymine in DNA and uracil in RNA, and vice versa. In order for DNA to make proteins, it must be transcribed by messenger

RNA (mRNA). The mRNA "reads" the DNA three bases at a time, matching its complementary bases to it. These groups of three bases are called codons, and each codon codes for a different amino acid. Chains of amino acids make up proteins. Therefore, it is vitally important that the DNA has the correct sequence of base pairs in order to make proteins correctly. A single point mutation could have no effect, or it could alter the protein that is produced and render it useless.

Point mutations are sometimes caused by mutations that spontaneously occur during DNA replication. The rate of mutations may also increase when a cell is exposed to mutagens, which are environmental factors that can change an organism's DNA. Some mutagens are X-rays, UV rays, extreme heat, or certain chemicals like benzene.

Types of Point Mutations

Substitution

A substitution mutation occurs when one base pair is substituted for another. For example, this would occur when one nucleotide containing cytosine is accidentally substituted for one containing guanine. There are three types of substitution mutations:

- Nonsense;
- Missense;
- Silent.

A nonsense mutation occurs when one nucleotide is substituted and this leads to the formation of a stop codon instead of a codon that code for an amino acid. A stop codon a certain sequence of bases (TAG, TAA, or TGA in DNA, and UAG, UAA, or UGA in RNA) that stops the production of the amino acid chain. It is always found at the end of the mRNA sequence when a protein is being produced, but if a substitution causes it to appear in another place, it will prematurely terminate the amino acid sequence and prevent the correct protein from being produced.

Like a nonsense mutation, a missense mutation occurs when one nucleotide is substituted and a different codon is formed; but this time, the codon that forms is not a stop codon. Instead, the codon produces a different amino acid in the sequence of amino acids. For example, if a missense substitution changes a codon from AAG to AGG, the amino acid arginine will be produced instead of lysine. A missense mutation is considered conservative if the amino acid formed via the mutation has similar properties to the one that was supposed to be formed instead. It is called non-conservative if the amino acid has different properties that structure and function of a protein.

In a silent mutation, a nucleotide is substituted but the same amino acid is produced anyway. This can occur because multiple codons can code for the same amino acid. For example, AAG and AAA both code for lysine, so if the G is changed to an A, the same amino acid will form and the protein will not be affected.

This image depicts the different types of substitution point mutations. It shows the DNA codon, the resulting RNA codon, and the amino acid produced.

	No mutation	Point mutations			
		Silent	Nonsense	Missense	
				conservative	non-conservative
DNA level	TTC	TTT	ATC	TCC	TGC
mRNA level	AAG	AAA	UAG	AGG	ACG
protein level	Lys	Lys	STOP	Arg	Thr

basic
polar

Insertion and Deletion

An insertion mutation occurs when an extra base pair is added to a sequence of bases. A deletion mutation is the opposite; it occurs when a base pair is deleted from a sequence. These two types of point mutations are grouped together because both of them can drastically affect the sequence of amino acids produced. With one or two bases added or deleted, all of the three-base codons change. This is called a frame shift mutation. For example, if a sequence of codons in DNA is normally CCT ATG TTT and an extra A is added between the two cytosine bases, the sequence will instead read CAC TAT GTT T. This completely changes the amino acids that would be produced, which in turn changes the structure and function of the resulting protein and can render it useless. Similarly, if one base was deleted, the sequence would also shift.

Examples of Diseases Caused by Point Mutations:

Cystic Fibrosis

Cystic Fibrosis (CF) is a recessive inherited disorder most common among people of European descent. In the United States, 1 in 3500 newborns is born with cystic fibrosis, and 1 in 30 Caucasian Americans is a carrier. There are many different mutations that can cause CF, but the most common one is a deletion of three nucleotides in the cystic fibrosis trans-membrane conductance regulator (*CFTR*) gene that results in the loss of the amino acid phenylalanine and causes an incorrectly folded protein. CF is associated with thick, sticky mucus in the lungs and trouble breathing, salty sweat, infertility in certain individuals, and a shortened life expectancy.

Sickle-cell Anemia

Sickle-cell anemia is a recessive disorder caused by a single substitution in the gene that creates hemoglobin, which carries oxygen in the blood. Normally, glutamic acid is produced in the chain, but the substitution causes valine to be produced at that spot instead. When people have two copies of this mutation, it results in thin sickle-shaped blood cells that sometimes cannot carry oxygen properly. About 80% of people with sickle-cell disease are in sub-Saharan Africa, where being a carrier for sickle-cell anemia (having only one copy of the gene, not two) actually helps protect against malaria. It is also found in other parts of the world such as India and the Middle East, and affects about 1 in 500 African Americans. Symptoms include anemia, obstruction of blood vessels, and chest pain, and it is treated with folic acid, blood transfusions, bone marrow transplants, and certain prescription drugs.

Tay-sachs

Tay-sachs disease is another recessive disorder caused by point mutations. Different mutations can cause this disorder, but they are all found on the *HEXA* gene on chromosome 15. Tay-Sachs causes nerve cells to deteriorate over time, which in turn results in the decline of physical and mental functioning. Both child and adult-onset forms of the disease occur, and children with the disease usually die before the age of four. About 1 in 320,000 newborns in the United States develop Tay-sachs. It occurs in higher frequencies in Ashkenazi Jews, Cajuns, and French Canadians (about 1 in 3500 in these populations), although the mutations associated with the disease are different in each population. There is currently no treatment or cure.

Frame-shift Mutations

Frame-shift mutations are insertions or deletions in the genome that are not in multiples of three nucleotides. They are a subset of insertion-deletion (indel) mutations that are specifically found in the coding sequence of polypeptides. Here the numbers of nucleotides that are added or removed from the coding sequence are *not* multiples of three. They can arise from extremely simple mutations such as the addition or removal of a single nucleotide.

Frame-shift mutations do not include substitutions where a nucleotide replaces another. In substitution mutations, the polypeptide only changes by a single amino acid. Frame-shift mutations also do not include indels in the non-coding or regulatory regions of the genome because these mutations do not have any direct effect on amino acid sequence, though protein regulation may change.

Effects of Frame-shift Mutations

Frame-shift mutations are among the most deleterious changes to the coding sequence of a protein. They are extremely likely to lead to large-scale changes to polypeptide length and chemical composition, resulting in a non-functional protein that often disrupts the biochemical processes of a cell. Frame-shift mutations can lead to a premature end to translation of the mRNA as well as the formation of an extended polypeptide.

The amino acid sequences downstream of the frame-shift mutation are also likely to be chemically distinct from the original sequence. For instance, if a frame-shift mutation occurs in an integral trans-membrane protein, it could vastly alter the stretch of hydrophobic residues that span the lipid bilayer making it impossible for the protein to be present in its subcellular location. When such errors occur, the cell often perceives the lack of functional protein and tries to compensate by up regulating the expression of the mutated gene. This can even overwhelm the translation machinery of the cell, result in a large number of misfolded proteins that could eventually lead to large-scale impairment of all functions of even cell death.

Diseases caused by frame-shift mutations in genes include Crohn's disease, cystic fibrosis, and some forms of cancer. On the other hand, when some proteins become dysfunctional, they could have a protective effect, as seen in the resistance to HIV in people with a chemokine receptor gene (CCR5) containing a frame-shift mutation.

Since frame-shift mutations are usually changes to the genetic material in every cell, it is rare to find a cure. Most interventions are palliative.

The Genetic Code

The core reason for the presence of frame-shift mutations is the body's mechanism for translating genetic information into amino acid sequences through a triplet-based genetic code. This means that every set of three nucleotides on an mRNA represents either an amino acid or an instruction to cease translation.

Discovery of the Genetic Code

Mendel's initial experiments on the transmission of genetic traits pointed towards a discrete physical and chemical entity that carried genetic information. Based on the bulk biochemical analysis of cells, four major components were detected – carbohydrates, fats, proteins and nucleic acids. Any of these components could represent genetic material.

Initial investigations into the chemical nature of the genome hypothesized that proteins, with 20 amino acids, were most likely to carry Mendel's factors or genes. However, later experiments indicated that nucleic acids were the carriers of genetic information. This presented an interesting difficulty. While nucleic acids had been analyzed chemically as being polymers made of 4 different nucleotides, it wasn't clear how the information for the dazzling variety of forms and functions in the body could arise from just 4 nucleotides.

Triplet Codon

A little later, the central dogma of molecular biology indicated that most organisms used RNA as the intermediate between DNA and proteins. This brought up the next question of how four bases could carry the information to encode 20 amino acids. If every nucleotide coded for a single amino acid, then only four amino acids could be reliably and reproducibly coded. If every two nucleotides encoded an amino acid, it would still lead to only 16 amino acids. Therefore, a minimum of three nucleotides was needed to code for 20 amino acids.

There are 64 permutations possible from nucleotide triplets where each position in the triplet can be one of 4 nucleotides. These nucleotide triplets were named codons. This also gave rise to the idea of redundancy – every amino acid could be represented by more than one codon triplet. Some experiments also revealed that codons were 'read' by the translation machinery as discrete chunks of 3 bases. That is, ribosomes 'see' these codons like a series of three-letter words. For instance, if an RNA molecule has the sequence AAAGGCAAG, then it can code for a maximum of 3 amino acids from the 3 codons AAG, GGC, and AAG.

Ribosome Translocation

The ribosome moves forward by three bases after each amino acid has been attached to the growing polypeptide chain. The way the ribosome moves is an important reason why frame-shift mutations are deleterious and have disproportionate effects on protein function. For instance, if the ribosome only moved by a single base each time, the previous mRNA containing 9 nucleotides can be read as AAA, AAG, AGG, GGC, GCA, CAA and AAG, giving rise to a polypeptide with 7 amino acids. If ribosome translocation only moved one base at a time, the insertion of a single nucleotide would only result in a small change to the amino acid sequence, and possibly no change at all to polynucleotide length.

Reading Frames

In the previous example, the polynucleotide chain can code for a maximum of 3 amino acids. However, depending on the upstream regions, the stretch cans also result in only 2 amino acids. That is, if the ribosome aligns with AAG or AGG instead of AAA initially, the nucleotide polymer is read in a different manner. This way, depending on the position of the translation start site, any coding sequence can be read in 3 different ways. Since most DNA is made of complementary double strands, it leads to a total of 6 different 'reading frames', only one of which results in the correct amino acid sequence for the final protein.

However, when there is an indel mutation, there is a shift in the reading frame downstream of the mutation. This results in a frame-shift mutation.

Examples of Frame-shift Mutation

The figure above shows the nucleotide and amino acid sequences in a wild type protein as well as the result of a nucleotide insertion, leading to the incorporation of incorrect amino acids and the premature end to polypeptide synthesis. While the original mRNA has a sequence of AUG AAG UUU GGC AUA GUG CCG, the insertion of an extra uracil residue at the ninth position changes the reading frame. Instead of producing a polypeptide of 7 amino acids beginning with methionine and continuing up to proline, it ends after 4 amino acids, with misincorporated leucine and alanine residues.

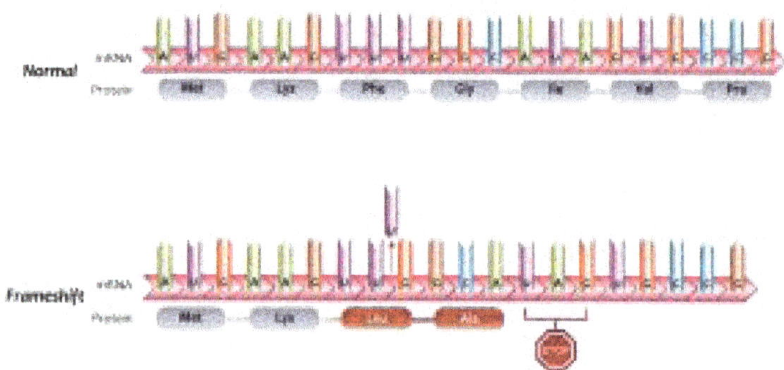

The image below shows the different types of mutations that could severely affect amino acid sequence. Panel A shows the substitution of 2 bases resulting in a premature stop codon, truncating the protein. Panels B and D demonstrate the effect of either the insertion of a single nucleotide or the deletion of 4 nucleotides. In both cases, a frame-shift mutation alters all downstream amino acid sequences. Panel C is a subset of indels where 3 (or multiples of 3) nucleotides are inserted or deleted. There is no frame-shift mutation. In this particular type of indel mutations, the number of nucleotides mutated is fairly low, there may be very limited effect on protein function as well.

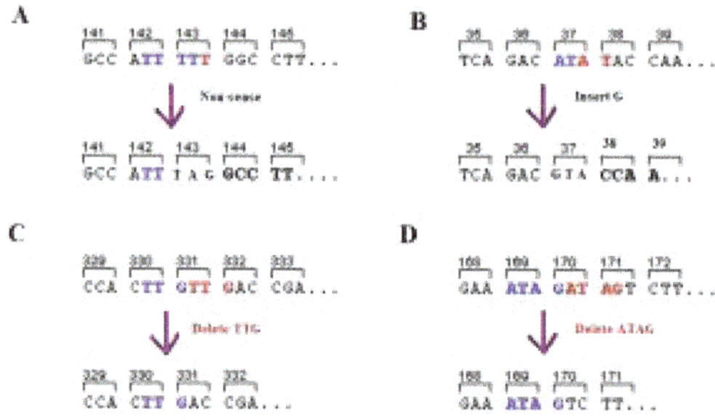

Synonymous Mutations

Synonymous mutations are point mutations, meaning they are just a miscopied DNA nucleotide that only changes one base pair in the RNA copy of the DNA. A codon in RNA is a set of three nucleotides that encode a specific amino acid. Most amino acids have several RNA codons that translate into that particular amino acid. Most of the time, if the third nucleotide is the one with the mutation, it will result in coding for the same amino acid. This is called a synonymous mutation because, like a synonym in grammar, the mutated codon has the same meaning as the original codon and therefore does not change the amino acid. If the amino acid does not change, then the protein is also unaffected.

Synonymous mutations do not change anything and no changes are made. That means they have no real role in the evolution of species since the gene or protein is not changed in any way. Synonymous mutations are actually fairly common, but since they have no effect, then they are not noticed.

Non-synonymous Mutations

Non-synonymous mutations have a much greater effect on an individual than a synonymous mutation. In a non-synonymous mutation, there is usually an insertion or deletion of a single nucleotide in the sequence during transcription when the messenger RNA is copying the DNA. This single missing or added nucleotide causes a frame-shift mutation which throws off the entire reading frame of the amino acid sequence and mixes up the codons. This usually does affect the amino acids that are coded for and change the resulting protein that is expressed. The severity of this kind of mutation depends on how early in the amino acid sequence it happens. If it happens near the beginning and the entire protein is changed, this could become a lethal mutation.

Another way a non-synonymous mutation can occur is if the point mutation changes the single nucleotide into a codon that does not translate into the same amino acid. A lot of times, the single amino acid change does not affect the protein very much and is still viable. However, if it happens early in the sequence and the codon is changed to translate into a stop signal, then the protein will not be made and it could cause serious consequences.

Sometimes non-synonymous mutations are actually positive changes. Natural selection may favor this new expression of the gene and the individual may have developed a favorable adaptation from the mutation. If that mutation occurs in the gametes, this adaptation will be passed down to the next generation of offspring. Non-synonymous mutations increase the diversity in the gene pool for natural selection to work on and drive evolution on a micro evolutionary level.

Impact of Mutation

Most often, mutations come to mind as the cause of various diseases. Though there are several such examples, disease-causing mutations are usually not very common in the general population.

Fragile X syndrome is caused by a dynamic mutation and occurs in 1 in 4,000 men and 1 in 8,000 women. Dynamic mutations are rather insidious since the severity of disease can increase as the number of nucleotide repeats increase. In those with fragile X syndrome, the nucleotide sequence CGG repeats more than 200 times within a gene called *FMR1* (for which the normal number is anywhere between 5 and 40 repeats). This high number of CGG repeats leads to delayed speech and language skills, some level of intellectual disability, anxiety and hyperactive behavior. However, in those with fewer numbers of repeats (55-200 repeats), most are considered to have normal intellect. Since the *FMR1* gene is on the X chromosome, this mutation is also heritable.

A variant of adult hemoglobin, known as hemoglobin S can occur due to a missense mutation, which causes the amino acid valine to take the place of glutamic acid. If one inherits the aberrant gene from both parents, it leads to a condition known as sickle cell disease. The disease gets its name from the fact that red blood cells, which are usually disc-shaped, contract and resemble a sickle. Those with the condition suffer from anemia, regular infections and pain. Estimates suggest that the condition occurs in 1 in 500 African Americans and about 1 in 1,000 to 1,400 Hispanic Americans.

Mutations can also occur due to environmental factors. For example, the UV rays from the sun, particularly UV-B waves, are responsible for causing mutations in a tumor suppressor gene called*p53*. The mutated *p53* gene has been implicated in skin cancer.

Mutations have other important implications. They create variation within the genes in a population. Genetic variants seen in more than 1 percent of a population are called polymorphisms. The different eye and hair colors, and the various blood groups that can occur, are all due to polymorphisms.

In the broad scheme of things, mutations can also function as tools of evolution, aiding in the development of new traits, characteristics, or species. "The accumulation of multiple mutations in a single pathway or in genes participating in a single developmental program are likely to be responsible for speciation.

References

- Molecular-evolution: iaszoology.com, Retrieved 25 March 2018
- Point-mutation: biologydictionary.net, Retrieved 19 July 2018
- Frameshift-mutation: biologydictionary.net, Retrieved 15 May 2018
- Synonymous-vs-nonsynonymous-mutations-1224600: thoughtco.com, Retrieved 28 April 2018
- Mutation-53369: livescience.com, Retrieved 18 July 2018

Permissions

All chapters in this book are published with permission under the Creative Commons Attribution Share Alike License or equivalent. Every chapter published in this book has been scrutinized by our experts. Their significance has been extensively debated. The topics covered herein carry significant information for a comprehensive understanding. They may even be implemented as practical applications or may be referred to as a beginning point for further studies.

We would like to thank the editorial team for lending their expertise to make the book truly unique. They have played a crucial role in the development of this book. Without their invaluable contributions this book wouldn't have been possible. They have made vital efforts to compile up to date information on the varied aspects of this subject to make this book a valuable addition to the collection of many professionals and students.

This book was conceptualized with the vision of imparting up-to-date and integrated information in this field. To ensure the same, a matchless editorial board was set up. Every individual on the board went through rigorous rounds of assessment to prove their worth. After which they invested a large part of their time researching and compiling the most relevant data for our readers.

The editorial board has been involved in producing this book since its inception. They have spent rigorous hours researching and exploring the diverse topics which have resulted in the successful publishing of this book. They have passed on their knowledge of decades through this book. To expedite this challenging task, the publisher supported the team at every step. A small team of assistant editors was also appointed to further simplify the editing procedure and attain best results for the readers.

Apart from the editorial board, the designing team has also invested a significant amount of their time in understanding the subject and creating the most relevant covers. They scrutinized every image to scout for the most suitable representation of the subject and create an appropriate cover for the book.

The publishing team has been an ardent support to the editorial, designing and production team. Their endless efforts to recruit the best for this project, has resulted in the accomplishment of this book. They are a veteran in the field of academics and their pool of knowledge is as vast as their experience in printing. Their expertise and guidance has proved useful at every step. Their uncompromising quality standards have made this book an exceptional effort. Their encouragement from time to time has been an inspiration for everyone.

The publisher and the editorial board hope that this book will prove to be a valuable piece of knowledge for students, practitioners and scholars across the globe.

Index

www.ingramcontent.com/pod-product-compliance
Lightning Source LLC
Chambersburg PA
CBHW061958190326
41458CB00009B/2900